History of Time: A Very Short Introduction

VERY SHORT INTRODUCTIONS are for anyone wanting a stimulating and accessible way in to a new subject. They are written by experts, and have been published in more than 25 languages worldwide.

The series began in 1995, and now represents a wide variety of topics in history, philosophy, religion, science, and the humanities. Over the next few years it will grow to a library of around 200 volumes – a Very Short Introduction to everything from ancient Egypt and Indian philosophy to conceptual art and cosmology.

Very Short Introductions available now:

For more information visit our web site

www.oup.co.uk/vsi/

Leofranc Holford-Strevens

THE HISTORY OF TIME

A Very Short Introduction

OXFORD
UNIVERSITY PRESS

OXFORD

UNIVERSITY PRESS

Great Clarendon Street, Oxford OX2 6DP

Oxford University Press is a department of the University of Oxford.
It furthers the University's objective of excellence in research, scholarship,
and education by publishing worldwide in

Oxford New York

Auckland Cape Town Dar es Salaam Hong Kong Karachi
Kuala Lumpur Madrid Melbourne Mexico City Nairobi
New Delhi Shanghai Taipei Toronto

With offices in

Argentina Austria Brazil Chile Czech Republic France Greece
Guatemala Hungary Italy Japan Poland Portugal Singapore
South Korea Switzerland Thailand Turkey Ukraine Vietnam

Oxford is a registered trade mark of Oxford University Press
in the UK and in certain other countries

Published in the United States
by Oxford University Press Inc., New York

© Leofranc Holford-Strevens 2005

The moral rights of the author have been asserted

Database right Oxford University Press (maker)

First published as a Very Short Introduction 2005

British Library Cataloguing in Publication Data

Data available

Library of Congress Cataloging in Publication Data

Data available

ISBN 978-0-19-280499-0

7 9 10 8 6

Typeset by RefineCatch Ltd, Bungay, Suffolk
Printed in Great Britain by
Ashford Colour Press Ltd, Gosport, Hampshire

Contents

Preface

The title of this book may suggest a survey of problems in philosophy or physics: whether time can have a beginning or an end; whether the laws of space–time cease altogether to apply in black holes; whether it would ever be possible to reverse the flow and change the past – a favourite fantasy with people who imagine that they alone would have the privilege of doing so, and forget that in the new improved past their parents might never have met.

These are indeed good questions, but no more my concern than the definition of time. About AD 268 the great Neoplatonist philosopher Plotinus observed that while we constantly talk about age and time as if we had a clear idea of what they were, when we investigate the question we find ourselves puzzled. The point was pithily restated some 130 years later by St Augustine: 'So what is time? If no one asks me, I know; if I seek to explain it, I do not.'

No pretence to greater wisdom is made in this book; whether time is a fourth dimension of the universe or a reified abstraction, whether it is continuous or atomistic, whether it can exist independently of motion to be measured, whether any meaning attaches to 'before' in the phrase 'before Creation' or 'before the Big Bang', are for others to determine. The same St Augustine, faced with the question what God was doing before he created the world, quoted, though he did not endorse, the jocular answer, 'Preparing hells for folk who invented clever conundrums like that'; I shall not take the chance that a true word was spoken in jest.

Nor shall I consider whether time proceeds in a straight line or in cycles. Although it is not true that linear time was a Judaeo-Christian speciality, set against the cyclical time symbolized in late Graeco-Roman paganism as a serpent devouring its tail, some philosophers did speak of time in cyclical terms. That poses conceptual problems that I shall not discuss; rather I shall confine myself to time in its ordinary-language or man-in-the-street sense, and shall concentrate on the methods by which its passage is and has been measured.

The English word 'time' may refer to a more or less closely defined period, from 'a short time', meaning not very long, to 'the time of the Pharaohs', some three thousand years; it may also refer to the 'indefinite continuous duration', as the *Oxford English Dictionary* expresses it, in which all events have taken place, are taking place, and will take place. This notion, the focus of Plotinus' and St Augustine's perplexity, presupposes a developed capacity for abstract thought; not only are various primitive peoples reported by anthropologists not to have such a concept of time, but in the epics ascribed to 'Homer' and dating from the 8th to 7th centuries BC that the Greeks regarded as the foundation of their culture, *chrónos* denotes only a lapse of time, not what we are tempted to think of as time itself. Nevertheless it already has that sense in the great Athenian lawgiver Solon of the early 6th century BC, who personifies it as a judge: 'in the court of Time'. Since then, this concept of indefinite continuous duration has been so familiar a concept to Western civilization that we find its absence unimaginable in any advanced culture; yet the case has recently been argued that neither the Hebrew Bible nor rabbinical literature displays it. However, in any but the simplest society, even if people are unaware of time as a thing in itself, they need to measure it. This book is about the methods by which the passage of time has been measured.

Homer has terms for years, months, and days; his references to disputes and lawsuits remind us of one important context for time-measurement, namely that even in his relatively simple society some cases must have turned, not on whether something had happened, but on whether it had happened before something else. If the two events had been witnessed by the same persons, there might be no problem; if not, both might be related to some third event, preferably one known to both parties and the judge, such as the local magnate's wedding. If there were no such event, difficulties would ensue unless the facts of the case could be plotted against a socially accepted measure of time.

The recording and coordination of human activities make it necessary to devise systems for relating events to a sequence of regular and predictable natural recurrences; since these systems were of artificial contrivance, and evolved in partial or complete independence one from another, they are different in many details. The range of variation, however, is limited by facts of nature, in particular the earth's rotation on its axis, the moon's revolution round the earth, and the earth's revolution round the sun; it is these that underlie the most widespread units for measuring time, the day, month, and year respectively.

The more complex life becomes, the more sophistication is demanded of the intellect not merely to distinguish one year, month, day, or subdivision of the day from another (the science of *time-measurement*), but to relate the years and so forth thus distinguished to each other (the science of *chronology*). This latter includes comparing the systems established for this purpose by different cultures to determine whether two apparently similar designations refer to two different things, or the same thing is lurking under two different names.

In much time-measurement fidelity to nature is in conflict with convenience; sometimes the former is sacrificed, as has repeatedly happened in Western methods of telling the time of day, sometimes the latter, as when Pope Gregory XIII made the Roman calendar more accurate but also more complex. By contrast, the designation of the year is free of natural considerations, being entirely a matter of convention; nevertheless, it is all too easily reified. In the early months of 1961 a manufacturer of electrical goods is said to have advertised its products in the name of a housewife called 'Mrs 1961', who because she was Mrs 1961 had to have the latest vacuum cleaner and the latest refrigerator. Her reward for thus increasing the company's sales was to disappear without trace in 1962.

Mrs 1961 was a victim of the delusion that years measured in our particular calendar and numbered in our particular era possess a reality beyond the conventions that created them. Yet in other calendars the year 1961 of the Christian era was not even a self-contained whole: in one Indian era it combined portions of 1882 and 1883, in another of 2017 and 2018, in Ethiopia of 1953 and 1954, in the Jewish calendar of 5721 and 5722, in the Muslim calendar of 1380 and 1381.

Such reification extends to larger units. 'The Sixties', meaning the 1960s, marks an entire decade as a time of political rebellion and cultural innovation; the 1890s (during which Oscar Wilde was convicted) are called 'the Naughty Nineties' because the elite chafed at the pretence of conforming to middle-class respectability. Centuries too are branded: 'in the 15th century religious devotion became increasingly personal and emotional', '18th-century English literature was dictated by the head and not the heart' – as if on the first day of 1401 or 1701 (not necessarily 1 January, as we shall see in Chapter 7), old ways of thought and feeling were abandoned like Mrs 1961's old vacuum cleaner.

When the emperor Trajan admonished Pliny, perhaps late in AD 110, that receipt of anonymous accusations was not compatible with 'our times', he meant quite specifically 'my reign', the principles by which he chose to rule. By contrast, modern journalists and politicians tell us that certain practices of government (though not that one) have no place in the 21st century, as if the date were a fact of nature and a legislator, so solidly is it reified. One purpose of this book is to combat such reification by illustrating the contingent and arbitrary nature of the measures to which it is applied.

Although the subject of this book is not politics or religion, I shall as occasion serves consider the political and religious implications in the choice of calendar, and the acceptance or rejection of reforms (e.g. the Gregorian calendar in Christendom, the 'Shahänshahi' era in Iran): even when the Government of India, in 1957, introduced a new secular calendar, it did not dare touch the multiplicity of religious calendars beyond substituting the tropical for the sidereal year. I shall also devote one chapter to a religious festival, the Christian Easter, not because of its religious significance but because of its calendrical complexity.

Nevertheless, my concern is with calendars as such rather than with their use or meaning; likewise, though much may be written about time as a social construct – and constructor – or about its perception by young and old, by men and women, or by office workers, factory hands, and peasants, there are others more qualified to write it.

Technical terms, when unavoidable, will be explained in a glossary; however, I note here that I have occasionally employed the single words 'feria', 'quantième', 'lune', and 'millésime' in place of the lengthier phrases 'day of the week', 'day of the month', 'day of the lunar month', and 'number of the year'. Numbers have been written in the scientific fashion, without commas: one thousand is 1000, ten thousand 10 000, one ten-thousandth 0.0001, one hundred-thousandth 0.000 01.

The traditional terms AD and BC have been retained, in preference to CE and BCE, for two reasons: adopting the latter causes the maximally distinguished BC 1 and 1 AD to become the minimally distinguished 1 BCE and 1 CE; and although, as a date for the birth of Jesus Christ the epoch is almost certainly wrong, it remains a commemoration of that event, and no other event of the same year can be proposed as an alternative of world significance. Attractive, especially in a globalized age, as a purely secular era may appear, the Christian era cannot be made secular by denying its origin.

List of illustrations

The publisher and the author apologize for any errors or omissions in the above list. If contacted they will be pleased to rectify these at the earliest opportunity.

Chapter 1
The day

Natural, artificial, civil day

The most fundamental unit of time-measurement is in most
societies the period of the earth's rotation on its axis, which is
normally known as the day. Unfortunately this word and its
equivalents in other languages are ambiguous: other meanings
apart, they may denote either the light period (daytime) as opposed
to the night, or the combination of daytime and night. In some
cultures, this combination is termed the night, as it used to be by
Celtic and Germanic peoples, who measured the length of journeys
or campaigns by the periods of inaction during darkness; this
practice – to which we still revert when booking a hotel – survives in
the English word 'fortnight', meaning 14 nights (formerly too in
'sennight', meaning a week). Nevertheless, the prevailing word is 'day'.

The two senses, 'daylight' and 'period of rotation', are distinguished
by the Latin author Censorinus, writing in AD 238, as *dies naturalis*
and *dies civilis* respectively; by the 7th century, however, educated
opinion had decided that the true day was the combined entity. As a
result, it was the latter that was called *dies naturalis*, the daytime
being renamed *dies artificialis*; accordingly Chaucer speaks of the
sun's 'artificial day' in the introduction to the *Man of Law's Tale*. It
is in this fashion that the terms 'natural' and 'artificial' day will be
used in this book.

In principle the natural day, being a segment of a continuum, may begin at any time. Some languages have an everyday word for a 24-hour period irrespective of starting point (e.g. Dutch *etmaal*, Russian *sutki*, Swedish *dygn*); this is particularly useful in measuring the duration of sea voyages, which unlike land journeys are not interrupted by nightfall. English has no corresponding term except the rare and scientific *nychthemeron*, a Greek word, literally meaning 'night-day', used by St Paul when he tells the Corinthians 'a night and a day have I been in the deep' (2 Cor. 11: 25). The New English Bible, anxious to avoid the implication that his ordeal began at sunset, renders 'for twenty-four hours'.

This unanchored natural day must be distinguished from the *civil day* in the strict sense, which is the natural day as reckoned from a particular point determined by law or custom. In the modern West, following Roman practice, and also in China, that point is midnight, but the Jewish and Muslim day is counted from sunset, as it was by the ancient Greeks and Babylonians; so (despite the midnight services that introduce Easter and Christmas) is the Christian liturgical day. The Egyptians (though not the Greeks of Egypt) reckoned from sunrise; in the same spirit most people in our own society, after midnight, call the next artificial day 'tomorrow' not 'today'. (In many languages, including English, the word for 'tomorrow' is related to that for 'morning', or is even the same, like Spanish *mañana*.) The peoples of ancient Umbria, however, began the day at noon, which struck the Romans as absurd. Noon was also the traditional beginning of the astronomical and nautical day, allowing all observations relating to a single night to fall on the same date; modern astronomers and sailors, however, have adopted the civil day.

Natural and social divisions

The apparent progress of the sun through the heavens can be measured, in the less cloudy climates, by observing the position or length of its shadow. It is recorded in the Bible that when, in the late

8th century BC, King Hezekiah of Judah fell ill, the prophet Isaiah induced a miraculous retreat of the sun's shadow by ten steps on an instrument evidently set up by the king's father, 'the steps of Ahaz'. Although the Authorized or King James Version speaks of 'degrees' on a 'dial' – meaning a sundial, not a clockface – the Hebrew word remains the same, *ma'ălôt*; more recent interpreters have supposed the steps to be a staircase or terrace, installed for use or beauty without regard to timekeeping. This would suit better with later midrashim, or elaborations of biblical stories, in which a scratch is made in the wall and a prophecy given that when the sun's shadow reaches the mark, such-and-such an event will take place.

These are not times of day as we understand them, any more than cock-crow or the natural and social events used as markers in Homer – 'when the early-born, rosy-fingered Dawn appeared', 'when the sun made his way towards ox-loosing', 'when a man rises for his supper after judging many disputes' – and long afterwards in the expositions of Jewish law known as the Mishnah; even midday and midnight are rather bands than points of time, halfway between sunrise and sunset or vice versa.

The hour

By contrast, the ancient Egyptians had for many centuries divided both the artificial day and the night into 12 'hours' each; in the former case, there was an earlier division into 10 hours of daytime plus 2 hours of half-light. The daytime hours were measured with shadow-clocks and sundials, those of the night identified by the succcessive risings of constellations. Every 10 days, a new constellation was recognized as rising with the sun (on each of the 9 succeeding days it rose 4 minutes earlier), yielding a set of 36 constellations known in Greek as *dekanoí*; this word, Anglicized as 'decan', was also used for an officer with 10 men under him, giving rise to our 'dean' and 'doyen'. For each 10-day period the decan that rose nearest to dawn, and the beginning of each hour, was noted in 'diagonal calendars', so called because each decan was one line higher from one column to the next (see Figure 1).

Such hours, technically called *unequal* or *seasonal* because they vary in length according to the time of year, were adopted by the Hellenistic Greeks and Romans (though the latter often divided the night into four *vigiliae*, or watches), and survived in normal use until the later Middle Ages. That is why Jesus, in St John's Gospel, asks 'Are there not twelve hours in the day?', meaning the artificial day. It is also why a midday rest is known as a *siesta*, Old Spanish for 'sixth', that is the sixth hour of the day (see box).

Ancient numbering of hours

- When it is said that on the day of the Crucifixion 'from the sixth hour there was darkness over all the land unto the ninth hour', this means from midday till mid-afternoon. Similarly, a Greek epigram states that there are 6 hours for working; the next 4 are for living, because the Greek letters zeta, eta, theta, iota, which were the normal notation for the numbers 7, 8, 9, and 10, spell the word *zêthi*, 'live!'

- The ecclesiastical offices of terce and nones owe their names to Latin *tertia* and *nona*, the 3rd and 9th hour respectively. However, a tendency to sing offices earlier than prescribed caused *noon*, the older form of 'nones', to mean 'midday'; the new sense is well established by the 14th century.

Although astronomers divided the natural day (reckoned from midday) into 24 equal, or equinoctial, hours, called by the latter name because at the equinoxes the nights and days are equal, other folk

1. Detail of Egyptian diagonal calendar

preferred the seasonal variety, which so long as work and travel were confined to daylight indicated both the time consumed and the time remaining. There were tables for converting the equal hours naturally measured by the clepsydra, or water-clock, into the unequal hours read off the sundial; not even the mechanical clock, which spread in Europe from the 14th century, gave immediate supremacy to equal hours, for the more complex clocks sometimes indicate unequal hours alongside the date and the position of sun and moon.

Once equal hours became the norm, however, it was more convenient to count them from midnight or midday than from sunrise or sunset; for that reason, people began to count the two sets of 12 hours before and after midday. Especially in English-speaking countries, this remains the norm outside bureaucratic and military usage, which favours the unambiguous count from 0 to 24. In Italy, however, there was a single sequence of 24 hours from sunset, the clock being adjusted from time to time as sunset moved later or earlier in the year; even now, when hours are counted from midnight, Italians freely use the 24-hour reckoning in everyday conversation. English-speakers do not arrange to meet for lunch 'at 13 hours' instead of 'one o'clock', meaning 1 p.m., but *alle tredici* is not in the least pretentious in Italian.

A variant on these 'Italian hours' characteristic of Majorca was a 24-hour sequence counted from dawn; these are known as 'Babylonian hours', from a false opinion in ancient authors that the Babylonian day began at sunrise. In fact, it began at sunset; night and artificial day were each divided into three 'watches', each in turn divided into four 'parts' or seasonal hours (see Figure 2); but the natural day was divided either (as we shall see) into 60ths, or into 12 *kaspu*, one for each sign of the zodiac, occasionally called *hôrai* in Greek, but commonly now known from a Eurocentric point of view as 'double hours'.

Double hours were adopted by the Chinese in 102 BC, displacing a previous division into 10 parts. The decree establishing the French

2. Babylonian ivory with calculation of length of hours

revolutionary calendar also envisaged a decimal division of the day into 10 hours, each of 100 minutes, themselves each of 100 seconds, to take effect on 1 vendémiaire year III (22 September 1794). Although the scheme would prove impractical, 10-hour clockfaces were made (see Figure 3).

Smaller divisions

The arithmetic of ancient Babylon was based on the number 60; accordingly, astronomers (despite the existence of double hours) divided the natural day into 60 parts, these parts in turn into 60ths, and so on. The length of the synodic month, for instance, was estimated at 29 days + 31/60 + 50/3600 + 8/216 000 + 20/ 12 960 000, which modern scholars write as 29;31,50,8,20 days.

3. French revolutionary clockface, showing division into 24 hours (outer ring) and 10 hours (inner ring)

Greek astronomers divided the natural day into 24 equinoctial hours, each of 15 *moîrai* or 'parts', the same word as they used for degrees of arc, since in either case there were 360 parts to the whole (Ptolemy, in the 2nd century AD, preferred to speak of 'equinoctial times'); we also find *stigmḗ*, 'point', used for half a *moîra*.

More complex systems are found in post-classical texts, both Greek and Latin (see box). Yet although small units were useful for astronomical and astrological purposes, or for displaying erudition, such conceptual division outran practical measurement, for there was no way of isolating atoms of time, 22 560 or 25 920 of the hour.

The neuter adjective *minutum*, 'tiny thing', was variously used for 1/15 hour (4 min.), 1/10 hour (6 min.), and 1/60 day (24 min.); but it never denoted 1/60 hour, which was an *ostentum*. In the later Middle Ages, however, we find a new sexagesimal division of the hour into *primae*, *secundae*, and *tertiae minutae* (*partes* understood). This system, already used for degrees of arc, has given rise to our 'minutes' and 'seconds'; 'third minutes', or 60ths of the second, abbreviated ‴, have largely given way to decimals.

Apparent and mean solar time

The triumph of the clock over the sundial as the preferred instrument for measuring time brought about a further change, besides the adoption of equinoctial hours: the displacement of the *apparent solar time* shown on the sundial by *mean solar time* shown on the clock. If the earth's orbit were a circle with the sun at its centre, and if its axis of rotation were perpendicular to that orbit, there would be no difference; but since the earth's orbit is elliptical, and the axis of rotation is inclined, the length of the natural day, measured against the 24 equinoctial hours of the clock, varies by half an hour over the course of the year. The difference between apparent and mean solar time is known as the *equation of time*; when apparent solar time is ahead of mean, the value is positive, when behind it is negative (see Figure 4).

Subdivisions of the hour

Byzantine Greek

1 hour = 5 *leptá* ('small things')	12 min.
1 *leptón* = 4 *stigmaí* ('points')	3 min.
1 *stigmḗ* = 2 *rhopaí* ('impulses')	1½ min.
= 3 *endeíxeis* ('showings')	1 min.
= 12 *rhipaí* ('blinks')	15 sec.
1 *rhipḗ* = 10 *átoma*	1½ sec.

Medieval Latin

1 hour = 4 *puncta* ('points')	15 min.
1 *punctum* = 2½ *minuta*	6 min.

OR

1 hour = 5 *puncta*	12 min.
1 *punctum* = 2 *minuta*	6 min.
1 *minutum* = 4 *momenta* ('impulses')	1½ min.
= 6 *ostenta* ('showings')	1 min.
1 *momentum* = 12 *unciae* ('ounces')	7½ sec.
1 *uncia* = 47 or 54 *atomi*	

A 7th-century Irish writer makes the *minutum* $2\frac{2}{3}$ moments
(4 min.); Hrabanus Maurus (9th century) calls this a *pars*
(cf. classical Greek *moîra*).

Hebrew

1 hour = 1080 *ḥălāqîm* ('parts', 'minims')	
1 *ḥaleq* = 76 *rəgā'îm* ('moments')	

For calendrical purposes, the *regá* is 1/82 080 hour; but a
Talmudic text declares that the *regá* for which God's indignation lasts is 1/58 888 hour.

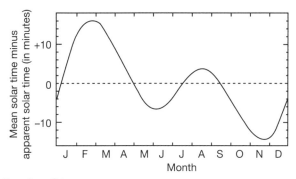

4. Equation of time

The standardization of time

Even when mean solar time had been adopted for purposes of legal definition (as in 1792 by Great Britain), it still varied with the local meridian: for every 15′ of longitude that one place lies to the east of another, the same nominal time arrives one minute earlier. So long as transport was confined to horse-drawn or water-borne traffic, and communication to the speed of horse or bird, that did not matter; but in the 19th century it made no sense that a train travelling at a given speed so many miles due west should appear to complete them sooner than one travelling at the same speed an equal number of miles due east, or that a telegraph message transmitted from east to west should appear to arrive before it had been sent.

Accordingly, the railway companies drew up their timetables, and set their station clocks, in accordance with Greenwich Mean Time, measured from a brass line at the Royal Observatory. Despite objections, not merely from certain civic dignitaries, but even from the Astronomer Royal, on the ground that if the clock said noon when the sun was not overhead it was lying, the new standardized time prevailed, and in 1880 was enshrined in statute. So completely has local time been forgotten that the tradition still observed at

Christ Church, Oxford, that one is not late till five minutes past the appointed time, that is to say till one is late by local mean solar time (longitude 1° 15′ W.) as well as Greenwich, is regarded even in other Oxford colleges as no more than an amiable eccentricity.

Time zones

Other countries similarly standardized their time; but while that was good enough for railway companies, it was not good enough for the international telegraph, which needed, if not a single time throughout the world, at least a single standard to which all local times could be referred. Since the globalization of trade and transport had made it convenient that all maps and charts should indicate the same longitudes in whichever country they were produced, as being so many degrees east or west of a prime meridian, this prime meridian would also yield the standard time.

In October 1884 an International Meridian Conference at Washington, DC adopted a US proposal that the prime meridian should be that 'passing through the centre of the transit instrument at the Observatory of Greenwich'; this has been the norm ever since, though for many years afterwards French maps continued to show 0° at Paris. (As a sop, the Conference adopted a French proposal for research into decimal measurement of angles and time.) In consequence, Greenwich time became the universal standard of reference, all other times being stated as so many hours ahead of or behind it. Again, the French proved resistant, ultimately succumbing only in 1911 and even then saving face by defining legal time as Paris Mean Time minus 9 minutes 21 seconds.

A few impractical extremists wished to make Greenwich Mean Time the universal civil time as well; but while the objections of science or superstition to a clock that read midday some ten minutes too soon or too late might be summarily overruled, it was not the same when at this nominal midday the sun was on the horizon, or the sky was pitch-black. Instead – since time of day, unlike season, is not affected by latitude – the globe is divided into vertical bands, known as time zones, running between the poles. Their boundaries are somewhat irregular (see Figure 5), owing to political factors such as state borders or the decision of Iceland to adopt Greenwich time and of France and Spain (though not Portugal) to be 1 hour ahead of it (' + 1' for short). Whereas India and China impose a single time on their entire territory (respectively + 5½ and + 8), other large countries have more than one zone; the prize goes to Russia with 11, from + 2 in Kaliningrad to + 12 in Anadyr'.

The International Date Line

As Jules Verne explains when Phileas Fogg, who thinks he has lost his bet to travel round the world in 80 days, is informed by Passepartout that the day is Saturday not Sunday, by going eastwards 'he went ahead of the sun, and in consequence the days diminished for him by four minutes for every degree', making 24 hours for 360 degrees; as a result, while he saw the sun cross the meridian 80 times, his fellow members of the Reform Club back in London had seen it do so only 79 times. Conversely, had he travelled westwards, he would have lost a day instead of gaining it.

An eastward-bound traveller crossing the meridian 180° east of Greenwich needs to give back the gained day, a westward-bound traveller to regain the lost day; ships therefore repeat the day when eastward bound and suppress a day when westward, and air travellers must set their calendar watches one day back in the former case and one day forwards in the latter. When the meridian runs through land or divides islands within a political entity, the change is made at a suitable point to east or west; this modified meridian is known as the International Date Line. In territories lying west of this line, the time of day is up to 12 hours (in a few places over 12 hours) in advance of Greenwich, in those to the east it is behind.

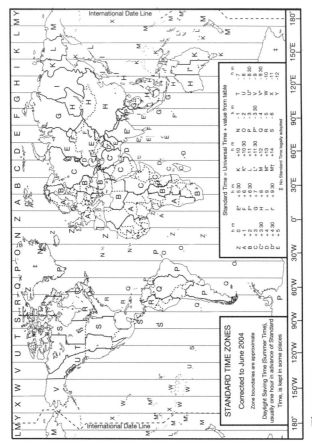

5. Time zones

Universal Time

For astronomical purposes, Greenwich Mean Time was reckoned until 1925 by the 24-hour clock commencing at noon; since then (in accordance with a hope expressed at the Washington conference) it has been reckoned from midnight, on a notional prime meridian a few metres away from the brass line. In 1928 it was renamed Universal Time, or UT. As observed, it is known as UT0; when corrected for the irregular movements of the terrestrial poles, or 'Chandler wobble', it becomes UT1. This is the astronomical and navigational standard.

Despite the existence of a further refinement, UT2, that attempted to correct for certain seasonal oscillations, it has proved impossible to base a uniform timescale on terrestrial movements. Once again, however, technology has outstripped nature: as the mechanical clock is more regular than the sun, so the atomic clock is more accurate than the earth. Accordingly, the second, for scientific purposes, is now defined as 'the duration of 9 192 631 770 cycles of radiation corresponding to the transition between two hyperfine levels of the ground state of the caesium-133 atom'; time as determined on this basis is known as TAI (*Temps atomique international*).

Since this takes no notice of the earth's rotation as measured by UT1, another standard for civil time, known as Coordinated Universal Time, or UTC, is employed, which is kept within 0.9 seconds of UT1 by leap seconds, either positive (23.59.60 added before 00.00.00) or negative (23.59.59 omitted; no actual case has yet occurred), at the end of June or December, at the behest of the International Earth Rotation and Reference Systems Service (IERS). UTC differs from TAI by an integral number of seconds; since January 2006 UTC − TAI = −33 seconds. In the English-speaking world it is often called Greenwich Mean Time, or GMT, although that term is also sometimes used for UT1.

Other measures of time are used by astronomers, such as Terrestrial Time (TT), 32.184 seconds ahead of TAI, which is used for calculating planetary positions in relation to the centre of the earth; the difference between this and UT1 is known as ΔT (delta T).

Daylight Saving (Summer Time)

In the early 20th century, the Chelsea builder William Willett advocated that during the spring and summer months clocks should be put forward so that people might enjoy the early morning light; having first suggested four successive 20-minute advances upon GMT, he finally settled for a single advance of one hour. The proposal languished until 1916, when in the interests of the wartime economy it was adopted by Germany and Austria-Hungary, the neutral Netherlands following suit. Great Britain did likewise for the duration of the war, and again in 1922, since when British Summer Time, or BST, has been in force for at least part of every year.

During the Second World War, this 'Summer Time' continued all winter from 1940 till 1945; from 1941 to 1945, and again in 1947 following a fuel shortage, a two-hour advance known as Double Summer Time was imposed for much of the year. An all-year BST, renamed 'British Standard Time', was enacted in 1968; but the persistence of darkness in the winter mornings (in northern Scotland as late as 10 a.m.) forced a reversion to GMT in 1971. By agreement within the European Union, Summer Time now begins at 1 a.m. GMT on the last Sunday in March and ends at that time on the last Sunday in October.

Most countries of the world observe daylight saving in some form, those south of the equator in different months from those in the north. In the United States, it runs by time zone from local 2 a.m. on the second Sunday in March to local 2 a.m. on the first Sunday in November; but states and US possessions may vote to exempt themselves, or (if they straddle time zones) parts of themselves;

thus it is not observed in Hawaii, the Hawaii-Aleutian time zone of Alaska, Arizona (except the Navajo Indian Reservation), or the territories of Puerto Rico, Virgin Islands, Guam, and American Samoa.

Chapter 2
Months and years

The period of the moon's revolution round the earth is at least notionally the *month*, called in most languages by a word meaning 'moon' or (as in English) a derivative of such a word; that of the earth's revolution round the sun is at least notionally the *year*. As we shall see, no calendar can do proper justice to both revolutions; either 'month' or 'year' must become an arbitrary measurement – just as the various measures called 'foot' may not match the length of any individual's bodily foot.

In some societies, days are grouped according to a second system running independently of months and years: the most widespread is the seven-day week. Many such systems are market-cycles; several such cycles of different lengths are found in Africa, but the most familiar are the ancient Roman eight-day *nundinum* and the French Revolutionary *décade* (see Chapter 5).

Years may be combined into centuries and millennia; longer groupings are found in India, understood as ages of the world, and formerly amongst the Eastern Maya, based not on the solar year or the 260-day cycle (see Chapter 6) but on the *tun* of 360 days and its multiples. The longer units have often been associated with the lifespan of the present world, which some people had imagined would end when the Christian era reached 2000; before that, a similar expectation had been attached in Russia to the 7000th year of the world, which in Western terms ran from 1 September 1491 to 31 August 1492.

The astronomical basis

The earth rotates on its axis; it also revolves round the sun, though to the naïve observer it is the sun that appears to revolve round the earth, just as the moon does. When sun and moon are close together, the moon cannot reflect the sun's light; the moment at which they have exactly the same celestial longitude is known as the conjunction or new moon, though the latter term is also applied to the first visibility of the moon after it has passed the point of conjunction. By contrast, when sun and moon face each other at 180°, we speak of opposition or full moon (Figure 6).

The sun's apparent revolution takes place along a path of sky known as the zodiac, from the Greek *zṓidia*, 'little animals', because it is divided into twelve 30° portions known as *signs* and called after constellations: the Ram (Aries), the Bull (Taurus), and so on (see Figure 7). However, these signs no longer correspond to the actual positions of the constellations; that is due to the slow circuit of the celestial sphere by the earth's pole over some 25 780 years, a circuit known as the precession of the equinoxes because it causes the

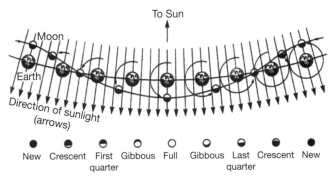

6. **The moon's phases**

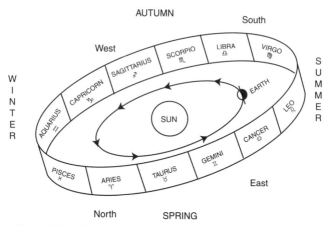

7. **Signs of the zodiac**

dynamical equinox – that intersection of ecliptic and celestial equator at which the sun's declination changes from south to north – to move slowly but steadily ahead relative to the constellations (see Figure 8). The First Point of Aries, which in the northern hemisphere marks the vernal equinox, is thus currently situated in the constellation Pisces, and is making its way towards Aquarius; but great civilizations existed while it was still in Taurus.

Most calendars are either lunar or solar. Lunar calendars are based, in theory, on the *synodic month* (from the Greek *súnodos*, 'conjunction'), or *lunation*, the period from new moon to new moon (though Tibet and northern India count from full moon to full moon), on average 29.530 59 days = 29 days 12 hours 44 minutes 2.976 seconds; 12 such months are grouped into a year. Solar calendars group days into years that measure the earth's revolution round the sun; these are subdivided into smaller units known as months but not governed by lunations. Most seek to match the tropical year (from the Greek *tropaí*, 'solstice'), the period of one complete revolution of the sun's mean longitude with respect to the

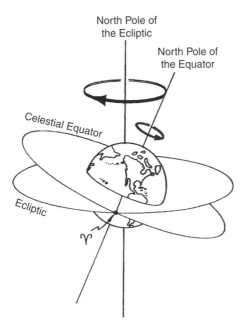

North Pole of
the Ecliptic

North Pole of
the Equator

Celestial Equator

Ecliptic

♈

8. The precession of the equinoxes

dynamical equinox; the current value is 365.242 19 days, or just
over 365 days 5 hours 48 minutes 45.2 seconds. If, however, the
intention is to mimic the period from vernal equinox to vernal
equinox, a closer average is 365.242 374 days, or just over 365 days
5 hours 49 minutes 1.1 seconds. These averages are gradually
changing; 2000 years ago they were 365.243 210 and 365.242 137
days respectively.

Owing to precession, the tropical year is somewhat shorter than the
sidereal year (from the Latin *sidus*, 'constellation'), or period
between two appearances in the same position relative to the stars,
which consists of 365.256 36 days or 365 days 6 hours 9 minutes
9.5 seconds. Most sophisticated calendars have been based on the
tropical year, except in India, which has a multitude of local

calendars both solar and lunar, the former based until 1957 on the sidereal year.

Unfortunately, 12 synodic months fall some 11 days short of a year, whether tropical or sidereal. For this reason, no calendar can be truly based on both; a choice has to be made. However, most lunar calendars attempt to keep the sun in view, whereas solar calendars pay no attention to the moon beyond the division into nominal months.

Lunar calendars

The oldest method of determining the new moon is by observation of first visibility: the competent authorities either watch the sky themselves or receive reports from persons deemed reliable. Although in principle this may seem the most accurate system, so long as the moon is considered as a visual rather than an astronomical phenomenon, it is open both to abuse in the interest of a faction and to interference from bad weather, which may be limited but not abolished by a rule that if after 29 days of the current month have elapsed the new moon is not observed on that evening it shall be deemed to be observed on the next, so that no month may contain more than 30 days.

Observation was also problematic in a community too large for rapid communication, and extremely inconvenient for astronomers who wished either to establish how many days had elapsed between two events in the past, or to predict the date of one in the future; for this reason, since the synodic month is a little over 29½ days long, a reasonably accurate lunar calendar will result from an alternation of 30-day 'full' and 29-day 'hollow' months, giving a year of 354 days; this schematic principle underlies the modern Jewish calendar (though with numerous complications) and the theoretical Muslim calendar used by astronomers and in conversion tables. It has the advantage that it can be extended as far ahead (or indeed back) as may be wished without regard to any external fact.

Intellectual advance offers another possibility, of calculating the conjunction, and beginning the month either on that day (as in China) or on the following day (as in southern India). In this way the real moon is still taken into account, as in the observed calendar, but relations between dates, whether past or future, can be established as precisely as in the purely schematic calendar, though only for so long as the calculations are deemed reliable.

Most lunar calendars attempt to correct the discrepancy between lunar and solar year by the addition, every few years, of an extra month; this is known as intercalation or embolism. It may take place when certain external conditions are met, as in the Jewish calendar while it was based on observation, and still does in Hindu lunar calendars. Alternatively, a rule may be applied; one rough method, known in antiquity and used in some early Christian Easter tables, is to add 3 months in 8 years, but more accurate is the addition of 8 months over a 19-year cycle. This is commonly known as the Metonic cycle, after Meton, a Greek astronomer who reportedly proposed it in 432 BC; however, it was first employed by the Babylonians, who possessed the most important lunar calendar of antiquity (see box). It is used in the modern Jewish calendar and (subject to certain qualifications) in the Chinese; it was also adopted by the Christian Church for calculating Easter (see Chapter 4).

Modern Westerners, to whom the Jewish and Muslim calendars are the most familiar non-solar varieties, distinguish lunisolar calendars with intercalation, which follow the moon but keep watch on the sun, from lunar calendars, which admit no intercalation and leave the sun out of account. However, the Muslim calendar, which is of this latter description, is exceptional; not only the Jewish, but the ancient Greek, Gaulish, Babylonian, and Chinese calendars are lunisolar, as are the moon-based calendars of India. It makes more sense to regard this as the predominant species of lunar calendar, and the non-intercalating calendar as the minority species rather than a third kind.

The Babylonian calendar

The Babylonian year, which began at the first new moon after the spring equinox, comprised 12 months, each beginning at the first sighting of the lunar crescent, called Nisanu, Aiaru, Simanu, Duzu, Abu, Ululu, Tashritu, Ahrasamnu, Kislimu, Tebetu, Shabatu, Addaru; days, which began in the evening, were counted forwards from 1 to 30 or 29. It is disputed at what point intercalation ceased to be *ad hoc* and was subjected to a 19-year cycle, but at least by the 4th century BC Addaru was repeated in years 3, 6, 8, 11, 14, and 19, and Ululu in year 17; new cycles began in 367/6 BC, 348/7 BC, 329/8 BC, 310/09 BC, and so on.

Within the month, the days may be numbered consecutively, as in the Jewish and nowadays in the Muslim calendar; but this is not the only system in existence. In several calendars the days of the waxing and waning moons are counted separately; Hindu lunar calendars conform to this pattern, as did (with complications) the ancient Gaulish calendar. In other calendars of this type, the age of the waning moon is counted backwards, so that the moon's appearance on any given day in the second half of the month is the mirror image of that on the corresponding day in the first; thus the 'tale of days', or numbering system, in a full month may run as follows:

1	2	3	4	5	6	7	8	9	10	11	12	13	14	15
15	14	13	12	11	10	9	8	7	6	5	4	3	2	1

In classical Arabic, dates were sometimes given according to this method (see Chapter 6). It is also found in medieval Europe, where it is called the 'custom of Bologna' (*consuetudo Bononiensis*), being favoured by that city's notaries; in addition, it is regular in lists of the two unlucky days in each month known as Egyptian days, of which the first is counted from the beginning of the month and the second from the end. In most cities of ancient Greece days were counted backwards after the 20th or 21st. Ancient Rome, as we shall see in Chapter 3, had a complex system of marker-days and backward counting.

Solar calendars

Although calendars exist in which the year begins when the sun reaches a fixed point in the sky (modern Iran, India since 1957) or enters a given constellation (India before 1957), it is far easier to work with the whole-number approximation of 365 days to the solar year. This is normally achieved by adding five 'epagomenal' or additional days to a term of 360 days divided into units conventionally known as months and regarded as the year proper, the extra days often being deemed unlucky.

That was the principle of the pre-Columbian calendars in Mesoamerica and of the Zoroastrian calendar still used by the Parsis; these are discussed in Chapter 5. It was also the principle of the ancient Egyptian civil calendar, the most venerable calendar of the ancient world, in which 12 months of 30 days were followed by 5 epagomenal days, known to the Egyptians as 'days upon the year'. Beside it there was a lunisolar ritual calendar, in which an embolismic month was added whenever the lunar year would otherwise have begun before the solar; in the 4th century BC an intercalation cycle was devised with 9 embolisms in 25 years. It was from this calendar that in the 6th century BC the months of the solar calendar took the names by which they were known to the Greeks and Romans.

Over all but the longest lifetime, the shortfall of the 365-day year is barely noticeable, but after a few centuries it is quite out of step with the seasons; it is therefore known as the *annus vagus*, or 'wandering year'. Notionally the calendar began with the heliacal rising of Sirius, whose Egyptian name sounded to the Greeks like 'Sothis'; this heralded the beginning of the Nile flood on which the country's life depended. In fact, however, compared with the actual rising it began almost one day earlier every four years (see Appendix A), or to put it another way the date on which Sirius rose was one day later; when once more the actual New Year coincided with the theoretical beginning, as happened on 20 July AD 139, there was great festivity.

By the 4th century BC, Greek astronomers were aware that the Egyptian year was too short; one of their number, Eudoxus, is said by the elder Pliny (d. AD 79) to have devised a 4-year cycle of solar years in which the first year was a leap year of 366 days. The underlying hypothesis was that the earth's revolution took 365 days 6 hours; that is in fact a little longer than the tropical year, and a little shorter than the sidereal year. In 238/7 BC the Macedonian king of Egypt, Ptolemy III, ordained that a sixth epagomenal day should be added every fourth year; the reform did not take effect, since the Egyptian priests were not going to admit an additional unlucky day at the behest of an alien ruler, and the Greek settlers, who would doubtless have complied, were not yet using the Egyptian civil year, but attempting to keep the Macedonian months in line with the local religious calendar.

The sixth epagomenal day was imposed once more by Imperator Caesar (soon to be given the name Augustus), when he had made himself king of Egypt in 30 BC. The exact details are controversial, especially since there are traces of calendrical experiment in the first years of his reign; but it seems that by 22 BC the reform had taken root in Alexandria (which technically was only 'by' Egypt, not part of it). In the rest of the country it took longer to be accepted; some astronomers, indeed, preferred the old calendar for its simplicity, since one could tell the number of days that had elapsed between any two observations recorded by date without worrying about leap days. However, it was the Alexandrian calendar – still used by the Coptic and (with different month names) the Ethiopian churches – on which the definitive Christian Easter reckoning would be based before it was translated into Roman terms (see Appendix B).

The solar cycle

Taken together, the 4-year leap-year cycle and the 7-day week yield a 28-year solar cycle, after which years begin on the same day of the week and occupy the same place in the leap-year cycle. This solar cycle, of high importance in the Easter computus, had by the 10th century become a calendrical unit in Iceland, where the year comprised an exact 52 weeks, or 364 days, an extra week being added 5 times in 28 years to make up the deficit (see Chapter 5).

Chapter 3

Prehistory and history of the modern calendar

The Roman Republican calendar

The calendar universally known and almost universally used today is a development of the Roman calendar as reformed by Julius Caesar in 46 BC and by Pope Gregory XIII in AD 1582. Before the first reform, it had been a debased lunisolar calendar with a common year of 355 days – one too many – for the sake of the odd number, which was thought to be more auspicious. For the same reason, there were no 30-day months as in normal lunar calendars; February had 28 days, but all the others an odd number, either 31 or 29.

Six of the months were named after the numbers 5–10, counting from March; but although there was a tradition that this had once been the beginning of the year and February the end (to which its shortness and its purificatory rituals gave support), in historical times the first month was January, named after Janus, the god of gateways who faced both backwards and forwards and who was named first in public prayers. The Romans themselves were aware of the contradiction, but found no more convincing answer than that King Romulus, the mythical founder of Rome, a soldier and a statesman but no intellectual, had not bothered to subdivide the bleak winter period between December and March.

Other cultures in Italy marked the beginnings of months and their mid-points at full moon; the Romans called these respectively *Kalendae* ('Kalends'), from an ancient verb *calare*, because the new moon had originally been announced on that day, and *Eidus* ('Ides'), from an Etruscan word meaning 'divide', which in the four 31-day months (March, May, Quinctilis, and October) fell on the 15th, in the rest on the 13th. But they also had a third marker-day, the *Nonae* ('Nones'), on the 7th of the 31-day months and the 5th of the rest, 8 days before the Ides as we should say, but as the Romans said the 9th day (*nonus* = 9th), since they preferred to count inclusively. (Christians still say that Christ rose on the 3rd day, counting Friday, Saturday, Sunday; compare the French *demain en huit* 'tomorrow week'.) All other days were named in relation to the next marker-day: the day before was called *pridie Kalendas/Nonas/Eidus*, the others 'the *n*th day before', once again counted inclusively, so that the day after the Kalends was the 6th before the Nones (*ante diem sextum Nonas*, abbreviated *a.d. VI Non.*) in the 31-day months, the 4th before them in the others, and so on.

Every day in the calendar was labelled with a letter from A to H, indicating its place in an eight-day market-cycle known by inclusive reckoning as the *nundinum*, or 'nine-day period'; once one knew the date of any market, the letter standing beside it would indicate all the other markets in the year. Other markings indicated whether the citizen assembly could meet, or the praetor hear lawsuits, on the given day; certain religious celebrations were also recorded, in accordance with tradition rather than current importance (see, for example, Figure 9).

In order to keep in line with the seasons, the body of priests known as the *pontifices* from time to time ordered the insertion of an extra 27-day month known as Interkalaris or Interkalarius, with Nones on the 5th and Ides on the 13th, either after 23 February (the Terminalia) or one day later; the remaining days of February were suppressed, so that the year contained 378 or 377 days. The

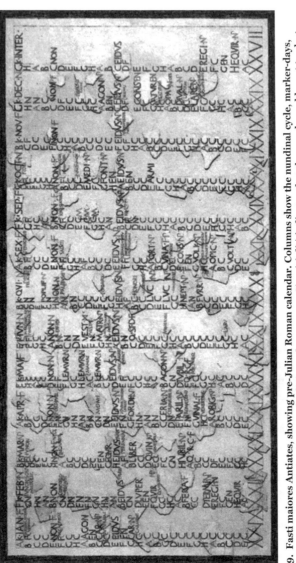

9. **Fasti maiores Antiates, showing pre-Julian Roman calendar. Columns show the nundinal cycle, marker-days, suitability for public business, and religious observances. C(*omitialis*) indicates that the people could meet to elect magistrates or vote on bills, F(*astus*) that the praetor could hear cases (all C days were also *fasti*), N(*efastus*) or NP (expansion unknown) that he could not, other markings that he could do so for only part of the day.**

nundinal cycle was interrupted at *K. Interk.*, but thereafter ran smoothly till the end of the year.

Intercalation was not ordered on scientific principles, but as political or other considerations might suggest (the *pontifices* were themselves politicians); sometimes the decision was taken so late that Cicero had to date his letters by days before Terminalia. The widespread superstition that extra days are unlucky caused the process to be suspended during the Second Punic War, so that the calendar raced some four months ahead of the sun; the solar eclipse of 14 March 190 BC was recorded on 11 July. When the *pontifices* resumed intercalation, they overcompensated to such an extent that when in 153 BC a military emergency compelled the new chief officers of state, or consuls, to enter office at once and embark on a spring campaign, they took over not on 15 March, as had been the norm since 222 BC, but on 1 January. Since that date was also the New Year, it was retained for the start of the consuls' term ever afterwards.

This was all the more convenient because the Romans regularly dated by the consuls; thus Julius Caesar was born 'on the fourth day before the Quinctilis Ides when Gaius Marius, for the sixth time, and Lucius Valerius Flaccus were consuls'. The conventional translation is 12 July 100 BC, though whether it was also 12 July in the retrojected Julian calendar (see below) we do not know.

The Julian reform

In 63 BC Caesar, at that point an ambitious young politician and not a military conqueror, was elected *pontifex maximus*, thus becoming responsible for intercalation; during the Gallic War it took place only in those years in which he could spend February away from Gaul, in 55 and 52 BC, and not at all in the ensuing Civil War. As a result, the calendar once more ran ahead of the sun, until Caesar, having defeated his implacable enemies, was free in 46 BC to take decisive action. Ordering not only a normal intercalation, but the insertion between November and December of two long months, together comprising 67 days, he extended that year – the last year of the confusion, as a late Roman writer calls it – to 445 days in order to make up for the missed intercalations of wartime.

From 45 BC onwards the new calendar was to be in force: the four 31-day months were unchanged, but the 29-day months were extended by either one extra day (April, June, September, November) or two (January, Sextilis, December), thus giving the year 365 days instead of 355. The position of Nones and Ides in these months was not affected; instead, the day after the Ides became the 18th or 19th before the Kalends instead of the 17th.

There was no further need of the intercalary month; but in order to keep the year in line with the seasons, a leap year was instituted in which 24 February was counted twice over. The extra day was known as *ante diem bis sextum Kalendas Martias*, the twice-sixth day before the March Kalends; from this comes the formal English term 'bissextile year' and the normal French *année bissextile* for 'leap year'.

Caesar decreed that the extra day should be inserted *quarto quoque anno*, meaning 'every fourth year', but since most Romans understood the phrase as 'every third year', after his murder a three-year cycle was instituted until Augustus, having himself become *pontifex maximus* in 12 BC, corrected the error by omitting intercalation after 9 BC (a leap year in both the correct and the faulty cycle), to be resumed in AD 8 and every four years thereafter. Since AD 8 was 52 years after 45 BC, this too must have been a leap year like the first year of Eudoxus' 4-year cycle, exactly as is presupposed in modern BC reckoning, which uses Caesar's calendar indefinitely retrojected. The first day of the Julian calendar was therefore the day we call 1 January 45 BC. That was a Friday, the day of Venus, governed by a beneficent planet from whose goddess Caesar's family claimed descent. (An alternative theory, that the calendar began on the 2nd with a common year, not only requires

Augustus to have made a mistake but overlooks the bad luck attaching to Saturday. Caesar's sense of public relations was far too good for that.)

In 44 BC the month of Caesar's birth, Quinctilis, was renamed Iulius to honour the deified dictator; in 8 BC, at the time of the correction, Sextilis, in which his heir had defeated the combined forces of Antony and Cleopatra, was renamed Augustus. Thereafter the calendar, despite a few linguistic changes such as *sexto Idus*, or later *Iduum*, for *ante diem sextum Idus* (the spelling *Eidus* had been given up), and the occasional but never permanent renaming of months after subsequent emperors, continued without further structural disturbance till 1582; neither the replacement of the *nundinum* by the week, nor successive changes in pagan and Christian feast days, affected the length and order of the months.

Although the Greek cities of the empire had their own calendars, the Roman calendar was also used as the empire-wide system of time-reckoning; but by the 5th century Greek-speakers had found the system of marker-days too cumbersome, and adopted the forward count from 1st to 31st, or whatever the last day was. This was much slower to be established in the Latin world, despite its use by Pope Gregory I (590–604); but it became the norm in the vernacular languages. In consequence, leap day was moved to 29 February (see box).

From Caesar to Gregory XIII

Caesar's reform presupposed a solar year of 365 days 6 hours, which is just over 11 minutes too long compared with the tropical year; the discrepancy amounts to 1 day in just over 128 years. From vernal equinox to vernal equinox it is less, 1 day in just over 131 years.

It was the vernal equinox by which the Christian churches defined Easter (see Chapter 4). The conventional Roman date, 25 March (established after the reform but derived from earlier Greek

Leap day

- Legally the additional day, *a.d. bis VI K. Mart.*, in Caesar's calendar fell on what we should call 25 February, following the normal *a.d. VI K. Mart.* on the 24th; however, unofficial usage reversed the order. This latter became the majority practice of the Western Church outside Norway and Iceland; it entailed postponing St Matthias' day in leap year from 24 February to the 25th, as is still the rule in the Roman Catholic Church. In the Church of England, however, postponement was given up in 1684, owing to an error by the then Archbishop of Canterbury that for political reasons was never put right.

- When dates are counted forward, without reference to the Roman marker-days, leap day is the 29th; that is the usage not merely of the civil calendar, but also of the Orthodox Church, in which 29 February is St Cassian's day.

astronomers, since it fits the 3rd century BC not the 1st), fell further and further behind the real equinox; in the 3rd century AD the true date was generally the 21st, which was adopted by the Alexandrian church in calculations that were to become definitive for all Christians, but in its turn became ever more obviously false.

From the 13th century onwards, when the discrepancy exceeded a week, proposals for reform were made, based on the principle that a certain number of days should be omitted as a one-off correction and thereafter intercalations should from time to time be suppressed, so as to reduce the excessive number of leap years. However, as we shall see in the next chapter, that would not have sufficed to correct the date on which Easter fell; on the one hand, it was the discrepancy between nominal and real equinox that gave the reformers their chief motive, on the other it was the change in Easter reckoning that generated the most hostility to their work.

In 1476 the great astronomer Johannes Müller, commonly known as Regiomontanus because he came from Königsberg in Bavaria, was summoned to Rome by Pope Sixtus IV for the purpose of reform, only to die soon after his arrival. Some forty years later Pope Leo X referred the question to the universities, which made no recommendation. Any hope of resolving the issue was dashed by the outbreak of the Reformation; a movement so hostile to clerical jurisdiction could not allow that popes had the right to reform a calendar laid down by Caesar, let alone for the sake of a feast not laid down by Scripture. Luther asserted that reform was the business not of the Church, but of the Christian princes; however, to avoid confusion, not least in the dates of fairs, they should act together or not at all. The advice was particularly pertinent for Germany, divided as it was into many states.

The Christian princes did nothing, and neither did the Council of Trent, which sat from 1545 to 1563 to consider reforms in the Roman Church. Indirectly, however, its final session reopened the question, by referring revision of the missal and the breviary – the Church's service-books – to the Pope. Although these revisions were duly executed by Pius V (whose breviary of 1568 included a botched revision of the Paschal lunar calendar), his successor Gregory XIII extrapolated a general power to reform the calendar on which the books were based; in 1578 he instituted a process that culminated four years later in the promulgation of the calendar that still bears his name.

In order that the equinox, currently on or about 11 March, should once again fall on the 21st, the Pope ordered that the day after 4 October 1582 should be called the 15th; further slippage was to be prevented by suppressing the leap day in centennial years (those ending in 00) unless exactly divisible by 400. That was not the most accurate correction, but it was the most convenient, being far easier to keep track of than alternative cycles; it also had the advantage that 1600 would still be a leap year, and the first suppression safely distant in 1700. This part of the reform is commonly known as the New Style.

Acceptance and rejection of the New Style

When Pope Gregory promulgated his Bull, not only was papal authority denied by Protestants, but opinion even in Roman Catholic countries laid a new emphasis on the rights of the civil power. Outside the Papal States it was the secular authorities that enacted the reform; the Italian cities and Spain were prompt to comply – so that St Teresa of Ávila died on 4 October 1582 and was buried on the 15th – but in France, where papal interference was widely resented, Henri III's desire to conform on the due date was thwarted (to the Pope's annoyance) by fierce opposition in the Parlement; not till December was the change made, when the 9th (marked by a solemn procession, led by the King, to pray for a royal heir) was followed immediately by the 20th. In some countries implementation took even longer.

The new calendar was much disliked both as being a change to age-old custom and as upsetting the traditional farmer's year and weather proverbs; worst of all, says a satirical German song, the Pope had interfered with St Urban's day (25 May), which was a predictor for the wine-harvest. If it was fine, meaning that the grapes would be plentiful, the peasants drank their fill of wine and poured some on the saint's statue; if it was wet, portending a poor harvest, they rolled his statue in the mire, or threw it into the river. The choice was between applying such observances ten calendar days later, or abandoning them altogether.

With the exception of two Dutch provinces, Holland (which omitted 16–24 December 1582) and Zeeland (which omitted 2–11 January 1583), the whole non-Catholic world retained the Julian calendar.

The Orthodox (and English high-churchmen) declared that the calendar of the '318 Holy Fathers at Nicaea' could not be altered except by an Oecumenical Council; Protestants who cared nothing for Nicaea objected that the equinox ought (as many Roman Catholics had expected) to have been made 25 March as in Caesar's day, and that the lunar tables were not perfectly accurate, but at bottom these were pretexts for refusing the work of the Pope. Even under Roman Catholic rule, the Greek possessions of Venice retained the Julian calendar, as did the Orthodox (and 'Uniats') of the Polish Commonwealth.

In England, Elizabeth I was intelligent enough to perceive the merits of reform and wise enough to take advice; her ministers were in favour, but her bishops would have none of it. The foremost mathematician and magus of the time, John Dee, calculating that 11 days and 53 minutes had crept in since Christ's birth, argued for omitting the last day of every month from January to September 1583 and the last two of October, so that England might set an example for the rest of Christendom, and even the Pope, to follow. This was the first of several schemes for English calendrical exceptionalism, and far sounder than the 592-year cycle propounded in 1621 by Thomas Lydiat on the basis of the Jewish rather than the Roman calendar, which won some fame but no following.

The new calendar made a little headway in the early 17th century, being accepted for example by the Duchy of Prussia and the Swiss canton of Wallis, but most Protestants were still using the Julian calendar as 1700 drew near, bringing with it an increase in the discrepancy from 10 to 11 days. This prospect induced the Dutch, German, Danish, and most Swiss Protestants to adopt the New Style (though not the Gregorian Easter; see Chapter 4). Sweden attempted to introduce it painlessly, by suppressing the 11 leap days from 1700 to 1740. That of 1700 was duly omitted, but not those of 1704 and 1708; as a result the decisive military defeat by Russia at Poltava, dated by British and Russian historians to 27 June 1709 Old Style, in most other countries on 8 July New Style, took place on

28 June Swedish style. Fear that this disaster was due to divine displeasure at interference with the calendar caused the Julian reckoning to be restored in 1712 by the addition of a 30th day to February.

A proposal to reform the English calendar had come to naught, not least at the anti-Popish urging of the mathematician John Wallis, who observed, amongst much else, that Easter could be found astronomically without changing the civil year – the very plan Sweden would adopt in 1740 – and that Scotland, still a separate kingdom, might not conform; since the Kirk refused to celebrate Easter as being unscriptural, it had no reason to care about the equinox. But once the need to persuade two Parliaments had been removed by the Union of 1707, several proposals were made for calendar reform, some quite radical, and advanced without regard to such practicalities as relations with the Continent.

The only sensible reform, adoption of the New Style, was eventually implemented by an Act of Parliament in 1751, which decreed the omission of the 11 days from 3 to 13 September 1752 (see Figure 10), and also switched the beginning of the English year from 25 March to 1 January as in Scotland and most other countries (see box, p. 40).

The Act distinguished between events such as Church festivals to be held on the 'nominal day' and those such as agricultural fairs, the determination of leases, or coming of age, postponed till the 'natural day': thus Michaelmas 1752 was to be 29 September New Style, but a Michaelmas fair would be held on 10 October. Several anomalies emerged, requiring additional legislation: at Chester, for example, where on the Friday after St Dennis's day (9 October) the mayoral ceremony was accompanied by a fair, the Act of 1751 required the former to take place 11 days before the latter. An amendment was hurriedly inserted into a bill on distemper in cattle.

September hath xix Days this Year. 1752

First Quarter,	*Saturday* the 15th, at 1 aftern.
Full Moon,	*Saturday* the 23d, at 1 aftern.
Laſt Quarter,	*Saturday* the 30th, at 2 aftern.

| 1 | f | Giles Abbot | 5 | 38 | 6 | 22 | ſecret | □ ♃ ☿ | 5 |
| 2 | g | London Burrt | 5 | 40 | 6 | 20 | memb. | Wind, | 6 |

Accoording to an Act of Parliament paſſed in the 24th Year of his Majeſty's Reign, and in the Year of our Lord 1751, the Old Style ceaſes here, and the New takes place; and conſequently the next Day, which in the Old Account would have been the 3d, is now to be called the 14th; ſo that all the intermediate nominal Days from the 2d to the 14th are omitted, or rather annihilated this Year; and the Month contains no more than 19 Days, as the Title at the Head expreſſes.

14	e	Holy Croſs	5	42	6	2	thighs	and ſtor-	7
15	f	Day decreaſ'd		45		2	hips	my Wea-	8
16	g	4 hours		46		18	knees	ther.	9
17	A	15 S. aft. Tri.		48		1	and	Fair and	10
18	b	Day br. 3. 45		50		14	hams	ſeaſonab.	11
19	c	Clo. flow 6 m.		52		12	'egs	☌ ☉ ☌	12
20	d	Ember Week		54		1	ancles	☌ ♀ ☿	13
21	e	St. Matthew,		56		8	feet	Rain and	14
22	f			56		6	toes	Windy.	15
23	g	Eq. D. & N.	5	58		4	head	☌ ☉ ☌	●
24	A	16 S. aft. Tri.	6	0		2	and		17
25	b	Day dec. 4, 34		2	6	c	face	□ ♃ ♀	18
26	c	S. Cyprian		4	5	58	neck	☌ ☉ ☿	19
27	d	Holy Rood		6		54	throat	Inclin. to	20
28	e	Clo. flow 9 m.		8		52	arms	☌ ☍ ☿	21
29	f	St. Michael		10		50	ſhould.	wet, with	22
30	g	St. Jerom		12		48	breaſt	Thunder.	☾

10. Almanac of 1752 showing September with 19 days

Which year was the longer?

Which year was the longer, 1751 or 1752, in (*a*) England, (*b*) Scotland, (*c*) France?

(*a*) 1752. In England, 1751 was the last year to begin officially on 25 March; it ended on 31 December, thus containing only 282 days.

(*b*) 1751. In Scotland, years had begun on 1 January since 1600, but the Old Style was still observed; hence, as in England, 11 days were removed from September 1752.

(*c*) 1752. In France, both reforms had already taken place, so that 1751 contained 365 days and 1752, being leap year, 366.

Late in 1752, a spectacularly uninhibited election campaign began for the two Oxfordshire seats, not contested since 1710, even though the poll would not be held till 1754. One of the 'New Interest' candidates (pro-government Whigs) challenging the established 'Old Interest' (oppositional Tories) was Lord Parker, the son of the Act's architect the Earl of Macclesfield. This made the calendar an issue in the early months of 1753; thereafter interest fell away, but the Oxfordshire excesses helped inspire Hogarth's satirical paintings of *The Election*. The first of these, *An Election Entertainment* (Figure 11), features a forlorn and trampled placard reading 'Give us our eleven days', which has given rise to the myth that there were riots against the New Style; riots indeed there were, but against the naturalization of Jews, which the government had just enacted and was intimidated into repealing.

In 1753 Sweden finally adopted the New Style, by omitting 18–28 February; the Protestants of the highly decentralized Grischun or Graubünden (not part of Switzerland till 1803) began

11. Hogarth, engraving of *An Election Entertainment*, showing placard 'Give us our eleven days' and hanged effigy of a Jew, alluding to the actual cause of riots

to fall into line in 1784, though the commune of Susch (in German Süs) had to be coerced by Napoleonic troops in 1811. From then on, the New Style was universal amongst Protestants as well as Roman Catholics.

Not till the 20th century did any Orthodox country accept the reform even for civil purposes; but the suppression in Bulgaria of 1–13 April 1916, followed by similar state action in Soviet Russia, Serbia, and Greece, provoked reconsideration. In May 1923 some churches agreed on a 'Revised Julian Calendar', comprising three reforms:

(i) the first 13 days of October 1923 should be omitted;
(ii) the only centennial leap years should be those that, when divided by 900, left a remainder of 200 or 600 (a far more accurate rule than the Gregorian);
(iii) the full moon on which the date of Easter depended should be determined not by the traditional rules (see Chapter 4), but according to the meridian of Jerusalem.

This last reform failed; of the other two, the first did not take place on time, and in many places has still not done so. Under state pressure, the Greek church – though not the self-governing communities on Mount Athos – accepted the reform from 10/23 March 1924; later that year state and church embraced it in Romania, but the Bulgarian church did not change till 1968, and the Russian, Serbian, Macedonian, and Georgian churches, along with the Orthodox in Jerusalem and Poland, still use the Julian calendar for the immovable feasts even though the civil dates are 13 days later. Christmas, which is celebrated in Greece on the same day as in the West, falls in these countries on 7 January; it thus follows the civil New Year, which as a non-religious festival was much promoted in the USSR to overshadow it.

The second reform (which some Westerners have wrongly credited to the Soviet regime) has met with widespread acceptance on paper;

but since 2000 was, and 2400 will be, a leap year in both calendars, not till 2800, a leap year in the Gregorian calendar but not the Revised Julian, will it be known whether the new rule has been put into practice.

Chapter 4
Easter

Easter, commemorating the Resurrection of Jesus Christ, is historically the most important of all Christian festivals, even though in some Western countries it has largely lost the religious significance it retains amongst the Orthodox; nevertheless it merits discussion in a broader context not only because it is often a public as well as a religious holiday, or indeed because even Christians may be baffled by its apparently capricious incidence, but because the history of its calculation illustrates many complexities of time-reckoning.

The origin of Easter is the Jewish Passover, known in Hebrew as *pesaḥ* but in Aramaic, the normal spoken language of Jews in Roman Palestine, as *pasḥā*, which in Greek became *páscha*. (From this comes our adjective 'Paschal'; the name 'Easter' is transferred from a Germanic festival.) Passover, in biblical times, was the slaughter of the Paschal lamb on 14 Nisan; it was eaten at nightfall, the beginning of the 15th by Jewish reckoning, and therefore of the seven-day Feast of Unleavened Bread.

According to St John's Gospel, the Crucifixion took place in the daytime of 14 Nisan; this date seems more plausible than the 15th offered by the other three gospels, since that day would have been profaned by the proceedings. It was therefore natural for early Christians, many of them still Jews, to commemorate the

Crucifixion on 14 Nisan, making Jesus the Lamb of God sacrificed to redeem humanity; the association was all the easier because in Greek *páscha* suggested the verb *páschein*, 'to suffer'. Even after Christianity had become a largely Gentile religion, the practice continued in Asia Minor, where the authority of St John was claimed for it, long after most other Christians had abandoned it. Such persistence was eventually classed as a heresy, known as Quartodecimanism.

However, since Jesus had risen on Sunday, that day was observed as a weekly feast day; in time it became normal, rather than to keep the anniversary of the Crucifixion, to celebrate that of the Resurrection, in the eastern provinces on the Sunday after Passover, elsewhere on the Sunday after a date found by independent calculation, both to avoid dependence on the Jews and because the increasingly widespread custom of fasting before it required knowing the date in advance. This entailed finding the 14th day of the first lunar month, called simply *tessareskaidekátē*, 'fourteenth', in Greek, but in Latin *luna quartadecima*, literally the 'fourteenth moon', abbreviated below as *luna XIV*. Once *luna XIV* had been found, the next Sunday had to be identified. The method of making these calculations is known as computus.

Easter limits

From the 3rd century onwards, most churches agreed that Easter should be the Sunday after *luna XIV*; but that was only the beginning. Two questions of principle remained: when the first lunar month began, and whether, should *luna XIV* be a Sunday, to observe Easter on that day or the following Sunday in order to keep apart from the Jews, who had not yet adopted a rule by which 14 Nisan could never be a Sunday. The latter became the more widespread practice; at Rome the conception of 14 Nisan as the first Good Friday caused Easter to be postponed even when *luna XIV* fell on Saturday, so that the feast was never earlier than the 16th lune or day of the lunar month (*luna XVI*), on which the Resurrection had taken place, and might be as late as the 22nd (*luna XXII*). On the other hand, it was essential that it should not fall after 21 April, the day of the Parilia, celebrating Romulus' foundation of the City, lest Christians, forced to fast on a civic feast, should be subject to hostility or temptation. Both the lunar limit of the 16th to 22nd lunes, and the lower solar limit of 21 April, are found in a calendar with base-year 222 inscribed in Greek on a stone chair now standing at the foot of the stairs in the Vatican Library.

In this calendar Easter may fall as early as 18 March; but Christians began to complain that Jews were no longer following their own rule that Passover should not precede the vernal equinox. This allegation was due partly to a misunderstanding of the Jewish rules, but partly to demonstrable variety in practice between one Jewish community and another (see Chapter 6). But if the Jewish Pascha ought not to precede the vernal equinox, it followed that the Christian Pascha ought not to do so either; by the 4th century the Church at Rome, where the date of the equinox was still taken to be 25 March, seems to have taken that view, though it could not be sustained in practice. At Alexandria the more rigorous principle was adopted that *luna XIV* itself, the proper date of Passover, must not precede the equinox, which was (more accurately) equated with 25 Phamenoth = 21 March; on the other hand, there was no lower limit, since 21 April (locally 26 Pharmouthi) was of no significance.

Early Easter cycles

There was also diversity, though not based on religious principle, in the methods used for finding *luna XIV*; many of them fell far short of the scientific standards already achieved by astronomy, suspect for its connection with astrology. The earliest method was the *octaeteris* or 8-year cycle, in which the lunar months were alternatively of 30 and 29 days, the civil leap day was counted as a lune in its own right, and an embolismic 30-day month was added in years 3, 6, and 8. This meant that 8 civil years exactly matched 8 lunar years, comprising 2922 days. The calendar on the Vatican chair was calculated ahead on this principle for 112 years. Unfortunately, since 99 actual lunations take not 2922 but just over 2923½ days, the table was soon found to be inaccurate; it was revised by a Latin writer, with new dates but on the same flawed principle, to run from 243.

Less inaccurate were 84-year tables, which had the advantage of comprising an exact number of weeks, so that each cycle began on the same feria as its predecessor. In the course of a cycle there were 31 embolisms, made whenever the epact or lunar age on 1 January exceeded 19; the leap day was not counted as a separate lune, and six lunes were omitted in order to keep lunar and solar cycle in order. This omission was called *saltus lunae*, 'the moon's leap', since the epact leapt over the intervening value, for example from 20 to 22; it took place either every 14 years, as in the table drawn up by one Augustalis, apparently in 3rd-century Africa (which permitted celebration on *luna XIV* and allowed Easter to fall on 16 March), or every 12 years up to year 72, as in the 4th-century Supputatio Romana, a shoddy piece of work that in some years gave two legitimate Easters, in others none.

The Alexandrian solution

At Alexandria the old 8-year cycle had long been banished in favour of the Metonic cycle of 19 years with 7 embolisms; although the precise course of development is uncertain, by 323 at the latest the computus had reached its final form (see Appendix B). It was based on a pseudo-Jewish calendar in which *luna XIV* ranged between the equinox on 25 Phamenoth (21 March) and 23 Pharmouthi (18 April), since celebration on that day was forbidden but not on the next, the earliest possible date for Easter was 26 Phamenoth (22 March), the latest 30 Pharmouthi (25 April).

In contrast to the Supputatio Romana, the Alexandrian cycle provided one and only one legitimate Easter each year; but since, unlike the 84-year table, it was not commensurate with the week, once the date of *luna XIV* had been determined it was necessary each time to find the next Sunday. This was done by calculating a variable known as the 'days of the gods', a pagan and astrological name that the Alexandrian church saw no need to change. Although Easter dates usually repeated at 95-year intervals, they did so fully only after 532 years.

It was the custom for the Patriarch of Alexandria to send out a circular letter every year announcing the date of Easter and making such comment on Church affairs as the times might demand. By the middle of the 4th century Patriarchs were claiming that the Council of Nicaea, called in 325 to discuss a troublesome heresy, had entrusted them with determining the date of Easter for Christians at large. This was not true, though immediately afterwards the Emperor Constantine had decreed that Christians must not 'celebrate with the Jews', that is be governed by the date of Passover; nevertheless, the Alexandrians gained increasing credence. By 360 the church in Milan was taking its Easter from Alexandria not Rome; in the early 5th century even Rome generally followed the Alexandrian date, provided this was no later than 21 April.

By the mid-6th century, Constantinople, which used a slightly different Metonic cycle, had adjusted the position of the *saltus* to yield the same Easter dates as the Alexandrian in all cases. The Armenian and both main Syriac-speaking churches accepted only part of the reform, so that in 4 years out of 532 they kept Easter on 13 April (Julian), while Constantinople kept the 6th. The first year of divergence was 570, the last 1824; the case is next due to arise in 2071. On occasions, the Greek and Armenian communities in Jerusalem have come to blows over the discrepancy.

Victorius and Dionysius

The Roman solar limit of 21 April was breached in 444, when Pope Leo I was persuaded to observe the 23rd, and in 455, with the greatest reluctance, the 24th; in consequence, the leading mathematician of the Latin-speaking world, Victorius of Aquitaine, was commissioned to draw up new Easter tables for papal use. His tables, extending over 532 years, purported to follow Alexandrian principles, discarding altogether the Roman 84-year cycle and solar limits; but he placed the *saltus* in the 6th year of the cycle instead of the 19th, let *luna XIV* range between 20 March and 17 April, gave different 'Latin' and 'Greek' dates when it fell on a Saturday, and refused to countenance Easter on 25 April.

The result was a mess: for 482 he gave a Latin date of 18 April on *luna XV* and a Greek date of 24 April (a Saturday!) on *luna XXII*, both lunes impossible for their respective churches. These tables would later attract scornful comment from all sides; nevertheless, they were widely used in the Latin-speaking world, because they appeared to give Easter dates for all time, and also retained the familiar reference to the epact and feria of 1 January.

At Rome, however, some dissatisfaction appears to have remained, for Victorius had abandoned Roman tradition without achieving unity with Alexandria or Constantinople. In 501, when Rome and Constantinople were in schism, Pope Symmachus celebrated on 25 March, according to the Supputatio Romana, and not the Alexandrian and Victorian date of 22 April; in 525, when the rift had been healed, the monk Dionysius Exiguus, invited to draw up a new table, made the final breach with Roman tradition by simply continuing the Alexandrian tables for a further 95 years from 532 to 626.

Dionysius' tables (see Figure 12) present five 19-year cycles, each set out in eight columns: years AD, indictions (see Chapter 6), epacts, concurrent days, lunar cycle, *luna XIV*, Easter, lune of Easter. Of these the last three are self-explanatory; the lunar cycle is that of Constantinople (three years behind the Alexandrian), added purely for comparison. The epacts are those of Alexandria, the concurrent days correspond to the 'days of the gods'; but neither term is explained, so that later users were obliged to work out the meaning

12. Sixth-century mosaic of Dionysius Exiguus' Easter tables. The sectors comprise columns showing: the year AD; the indiction; the epact; the concurrent; the position in the Byzantine lunar cycle; date of *luna XIV*; date of Easter; lune of Easter.

for themselves. Indeed, it is not apparent that Dionysius himself knew what they were: he could extrapolate their values easily enough, and as a practical man probably looked no further. As might be expected from one whose other main achievement, a compilation of church law, required learning and diligence rather than insight, his tables are accurate, his expositions defective.

Despite being commissioned, Dionysius' tables took over a century to oust Victorius' at Rome, and even longer elsewhere. Victorius' tables were perpetual, followed the traditional form, and appeared to be comprehensible (it was on close examination that the flaws appeared); Dionysius' would need recalculating after 95 years, ignored 1 January, and were difficult to understand. By Isidore of Seville's time, in the early 7th century, it had been discovered empirically that the epact corresponded to the lune of 22 March; in the early 8th century the Venerable Bede stated what Irish computists already knew, that the concurrent days (or, as we usually say in English, the 'concurrent') matched the feria of 24 March. Ever afterwards these explanations have been standard in Western writing, and all too often are supposed to be Alexandrian.

Insular Easter

In the British Isles, Easter was calculated on entirely different principles, from a table known by an irregularly formed Latin word as the Latercus and attributed to Sulpicius Severus in the mid-4th century. It was based on an 84-year cycle with *saltus* at 14-year intervals; the solar limits were 26 March for the earliest Easter and 23 April for the last, the lunar limits *lunae XIV* to *XX*, so that when *luna XIV* fell on a Sunday, that was Easter. When the rest of Christendom found out, it was scandalized.

The upper solar limit was the day after the Latin equinox; the other limits were taken from Augustalis. The novelty was the lunar calendar, which abandoned the alternation of full and hollow months: the full month ending in January was followed by three successive hollow months; the remaining lunar months each contained one day fewer than the solar months in which they ended. A careful choice of initial epact ensured that, like the Alexandrian computus, and unlike either Augustalis or the Supputatio Romana, the Latercus gave a single Easter date for every year.

When at the end of the 6th century St Colmán, or Columbanus, of Bangor (Co. Down) left Ireland for Gaul, his Latercus offended the local churches, which were committed to Victorius. In a letter to Pope Gregory the Great that veils insolence in humility, Columbanus defends his practice, declaring that the learned among his countrymen had rejected Victorius' tables as worthy rather of pity than of scorn (a judgement more charitable than those of Dionysius and Bede); that celebration on *luna XXI*, let alone *XXII*, is improper since the moon ought not to rise later than midnight on a feast celebrating the triumph of light over darkness; and that celebration on the same day as the Jews is no problem because Pascha belongs not to them but to the Lord (Exod. 12: 11).

Meanwhile, the papal mission to England introduced Roman practice into southern England, and came into conflict over this and other matters with the Britons; during the 7th century both Victorius' and Dionysius' tables made some headway in Ireland, above all in the south (though memory of the Latercus epacts survived in West Cork till the mid-19th century), but the monastery of Iona, founded by Colm Cille (St Columba), and those associated with it, remained steadfast for the Latercus, which missionaries brought thence to Northumbria.

The difference between the two practices caused difficulties in the Northumbrian court, where the Irish-educated King Oswy had married the Kentish princess Eanflæd, brought up in the Roman tradition; Bede reports that sometimes the King would be celebrating Easter while the Queen was still observing Palm Sunday. Until the details of the Latercus were known, this was thought to be an occasional event, due to his keeping *luna XIV*; in fact, it happened in at least half the years of their marriage before the question was resolved, but never for that reason. The discrepancies were mostly due to the differences between their lunar calendars, sometimes to those between their solar limits; in no year was the same day *luna XIV* for both, whether on Sunday or any other feria (see box).

By 664 King Oswy, and the clergy of his realm, had had enough. A

King Oswy's Insular and Queen Eanflæd's Roman Easter

M = March, A = April

Year	Insular Easter	Lune	Roman Easter	Lune Vict.	Dion.	Year	Insular Easter	Lune	Roman Easter	Lune Vict.	Dion.
643	6A	XV	13A	XIX	XVIII	653	14A	XIV	21A	XVIII	XVII
644	28M	XVII	4A	XXI	XX	654	6A	XVII	13A	XXI	XX
645	17A	XVIII	17A	XV Lat.		655	29M	XX	29M	XVII	XVI
			(24A	XXII Gr.)		656	17A	XX	17A	XVII	XVI
			24A		XXI	657	2A	XVI	9A	XX	XIX
646	2A	XIV	9A	XVIII	XVIII	658	22A	XVIII	25M	XVI	XV
647	22A	XVI	1A	XXI	XXI	659	14A	XX	14A	XVII	XVI
648	13A	XVIII	20A	XXI	XXI	660	29M	XV	5A	XIX	XVIII
649	29M	XIV	5A	XVII	XVII	661	18A	XVII	28M	XXII	XXI
650	18A	XVI	28M	XX	XX	662	10A	XX	10A	XVI	XV
651	10A	XVII	17A	XXI	XXI	663	26M	XVI	2A	XIX	XVIII
652	1A	XX	1A	XVII	XVI	664	14A	XVI	21A	XIX	XVII

synod was summoned at Whitby to determine whether Celtic or Roman tradition should be followed, both in Easter computus and in clerical tonsure; for whereas the Romans shaved the crown of the head, the Celts shaved everything from forehead to ears.

The Celtic cause was argued by Bishop Colmán of Lindisfarne, the Roman by Wilfrid, later Bishop of York, who dominated the debate, bullying poor Colmán, insulting St Columba, misrepresenting facts, but clinching the argument with fantastic assertions about the practice of St Peter. King Oswy opted for Rome, giving as ground that it was St Peter, not St Columba, who held the keys of heaven. Gradually the rest of the British Isles fell in line with Rome; the last to conform were the Welsh.

Although the struggle was described as being between the 19-year cycle (Victorian or Dionysian) and the 84-year cycle of the Celts, the mode of calculation was a detail; it was the difference in limits that raised questions of religion. The rivalry between the two versions of the 19-year cycle remained on the technical level; the decisive blow was struck by Bede in his *De temporum ratione* of 725, which did for Dionysius what Dionysius had failed to do (and Victorius had done for his own system) by setting out a complete exposition of his Easter principles together with a 532-year Easter table that did away with the need for recalculation every 95 years (see Figure 13). Within a century, Bede's table was in use throughout the West, even in Gaul, where Victorius died hard.

The lunar calendar

Although the Alexandrian Easter computus, like others, is based on a lunar calendar, the Alexandrians were not particularly concerned to know the lune of every day; in the West, by contrast, such knowledge was deemed so important that in monasteries the lune of the day would be announced along with its martyrs at prime. This entailed recalculating the new moons: it was not enough to convert Alexandrian dates into Julian, for though in both Alexandrian and Western practice the odd months of the lunar year were full and the even hollow, their incidence differed.

In Alexandria, the full months normally began in the odd-numbered solar months, but in the West, where the lunar month was identified with the solar month in which it ended, they normally began in the even-numbered months and finished in the odd. This was expressed by the versified rule *impar luna pari, par fiet in impare mense*, as it were 'odd moon in even month, even in odd'. That is to say, the moon has an odd number of days (29) in an even-numbered month, an even number of days (30) in an odd-numbered month.

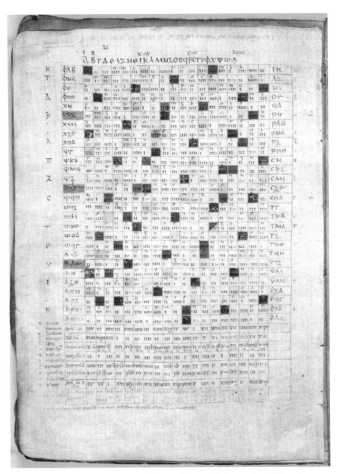

13. Bede's perpetual Easter table. Each row of the main table relates to the 19-year cycle beginning in the year AD indicated on the left with Greek numerals from 532 to 1045; the squares give the concurrent with marks for leap years and references to a table of lunar ages; shadings show 1st and 8th indictions from 538 and 545 onwards, and every 28th year from 532 onwards; the right-hand column gives the date of the last year in the previous Paschal cycle, from 18 to 531. The surrounding rows and columns give other miscellaneous information.

Since in the 7th century the leading experts on the computus were the Irish (from whose writings Bede learnt much, though he never says so), it is not surprising to find a treatise entitled *De ratione conputandi* that sets out a complete lunar calendar for the year on Dionysiac principles. However, though much of this text was to be copied into a book owned three hundred years later by St Dunstan, it was Bede's rather different system that became definitive. By careful placement of his embolisms he ensured that only in three years was the regular sequence of monthly epacts disrupted, and only in those years were some lunar months named after the solar months in which they began. Thereafter, when men as learned and ingenious as Hrabanus Maurus (*c.* 780–856) and Abbo of Fleury (*c.* 945–1004) produced new methods of calculating the same results, it was Bede whom they were refining.

In Bede's time, the lune of the day was found by adding to the epact (see box) parameters known as 'lunar regulars' to obtain the moon's age on the 1st of each month: thus for epact 11, the lune of 1 March was $11 + 9 = 20$, of 1 April $11 + 10 = 21$. From the 12th century, however, it was found directly from the so-called Golden Number, written in calendars against the dates of the new moons in the corresponding year of the 19-year cycle; similarly Bede's method of finding feriae from the concurrent with the aid of 'solar regulars' (for example, if the concurrent was 2, then the feria of 1 March was $2 + 5 = 7$, of 1 April $2 + 1 = 3$) was replaced in the later Middle Ages by the Sunday Letter (see Chapter 5).

The need for reform

However well Bede had done his work, he could not cure the basic faults of the solar and lunar calendars, namely that neither was accurate enough. By the 13th century, the solar calendar was far enough behind the sun for reforms to be proposed. This was no longer a task for an intelligent heir of Caesar and Augustus: in a

Epacts, Golden Numbers, concurrents

To find the Julian epact, divide the year AD by 19; if the remainder is 0, that is the epact, otherwise multiply it by 11, divide by 30, and the new remainder is the epact.

To find the Golden (i.e. important) Number, also called the prime (from *luna prima*), add 1 to the year, divide by 19, and take the remainder; if this is 0, the Golden Number is 19.

To find the concurrent (used only in the Julian calendar), add to the year AD its fourth part (ignoring fractions) and the parameter 4, divide by 7, and take the remainder; if there is no remainder the concurrent is 7, and 24 March is a Saturday.

Christian Europe reform touched on the far more ticklish problem of Easter.

In 1538 the Golden Number was 19, indicating *luna XIV* on 17 April; since this was a Wednesday, Easter fell on the 21st. Martin Luther observed that, according to the rule of the first Sunday after the first full moon of spring, it ought to have fallen on 17 March, which of course preceded the ecclesiastically defined equinox on the 21st. Had the New Style already been in force, but unaccompanied by any further reform, that day would have been called the 27th; however, it would not have been Easter because the Golden Number would still have indicated *luna XIV* on 17 April. Since this would now have been a Saturday, Easter would have fallen on 18 April, corresponding to the 8th Old Style and nearer new moon than full.

Moreover, to recalculate the lunar epacts – for example, reducing that of Golden Number 19 from 18 to 11, so that *luna XIV* became 25 March and Easter 1538 the 27th – would have been only a short-term solution, for the lunar calendar too was wrong (as Roger of Hereford had observed in the 12th century). Over 76 years, leap days included, it comprised 27 759 days; the corresponding 940 synodic months comprise (on average) 27 758.754 6 days, putting the moon some 5 hours 53 minutes 23 seconds ahead of the calendar. This discrepancy would reach a day every 310 years, making the epacts wrong again. A more fundamental reform was needed; it was enacted by Gregory XIII.

The Gregorian calendar

Pope Gregory's reform had three objectives: to restore the date of the vernal equinox, 21 March, universally though wrongly believed to have been decreed at Nicaea; to keep it there in future; and to reconcile the Easter calendar, so far as possible, with the moon. The first two were achieved by the New Style discussed in Chapter 2; the third objective, reform of the lunar calendar, was met by substituting for the simple Alexandrian 19-year cycle a highly complicated system based on the epact of the year, redefined as the lune of 31 December preceding (see Figure 14).

At each centennial common year the epacts are diminished by 1, but every 300 years they are increased by 1 save that every eighth such increase is delayed by a century; when both rules apply to the same year there is no change. These adjustments slightly over-correct the discrepancy between lunar calendar and moon. It is the epact, not as in the Julian calendar the Golden Number, that is used to find the lune of any given day. (The necessary tables will be found in the *Oxford Companion to the Year*, pp. 825–8.)

The reformers did not specify the longitude by which they calculated their new moons; this was necessary because they had often deliberately placed them a day or two late in order to avoid coincidence with Passover. This in Jewish usage no longer meant

TABVLA FESTORVM

Anni Domini	Aur. Num.	Epactæ	Lit. Diiicales Calend. Greg.		Plenilunia media Cal Gregor. D. H.	Lunæ xiiij. Calendarij Gregoriani	Septuagesima	Dies Cinerum	Pascha Calend. novi	Ascēfio Domini
3664	17	xix	f	e	25. 2.M	25. M	27 Ian.	13. Feb.	30. M	8.Maij
3665	18	*		d	13. 0.A	13. A	15. Feb.	4.Mar.	19. A	28.Maij
3666	19	xj		c	2. 8.A	2. A	31. Ian.	17. Feb.	4. A	13.Maij
3667	1	xxiij		b	22.17.M	21 M	23. Ian.	9.Feb.	27. M	5.Maij
3668	2	iiij	A	g	9.15.A	9. A	12. Feb.	29.Feb.	15. A	24.Maij
3669	3	xv		f	29.23.M	29. M	27. Ian.	13. Feb.	31. M	9.Maij
3670	4	xxvj		e	17.21.A	17. A	16. Feb.	5.Mar.	20. A	29.Maij
3671	5	vij		d	7. 6.A	7. A	8.Feb.	25.Feb.	12. A	21.Maij
3672	6	xviij	c	b	26.15.M	26 M	24. Ian.	10. Feb.	27. M	5.Maij
3673	7	xxix		A	14.12.A	14 A	12.Feb.	1 Mar.	16. A	25.Maij
3674	8	*		g	3.21.A	3. A	4.Feb.	21.Feb.	8. A	17.Maij
3675	9	xxj		f	24. 6.M	23. M	20.Feb.	6.Feb.	24. M	2 Maij
3676	10	ij	e	d	11. 3.A	11 A	9.Feb.	26.Feb.	12. A	21.Maij
3677	11	xiij		c	31.12.M	31. M	31. Ian.	17. Feb.	4. A	13.Maij
3678	12	xxiiij		b	19.10.A	18. A	20.Feb.	9 Mar.	24. M	2.Iunij
3679	13	v		A	8.19.A	8. A	5. Feb.	22. Feb.	9. A	18.Maij
3680	14	xvj	g	f	28. 3.M	28. M	28. Ian.	14. Feb.	31. M	9.Maij
3681	15	xxvij		e	16. 1.A	16. A	16.Feb.	5.Mar.	20. A	29.Maij
3682	16	viij		d	5.10.A	5. A	8. Feb.	25.Feb.	12. A	21.Maij
3683	17	xix		c	25.19.M	25 M	24 Ian.	10.Feb.	28. M	6.Maij
3684	18	*	b	A	12.16.A	13 A	13.Feb.	1.Mar.	16. A	25.Maij
3685	19	xj		g	2. 1.A	2. A	4.Feb.	21.Feb.	8. A	17.Maij
3686	1	xxiij		f	22.10.M	21 M	20. Ian.	6.Feb.	24. M	2.Maij
3687	2	iiij		e	10. 7.A	9. A	10.Feb.	27. Feb.	13. A	22.Maij
3688	3	xv	d	c	29.16.M	29. M	1 Feb.	18. Feb.	4. A	13.Maij
3689	4	xxvj		b	17.14.A	17 A	20.Feb.	9.Mar.	24. A	2.Iunij
3690	5	vij		A	6.22.A	6 A	5. Feb.	22.Feb.	9. A	18.Maij
3691	6	xviij		g	27. 7.A	26 A	28. Ian.	14.Feb.	1. A	10.Maij
3692	7	xxix	f	e	14. 5.A	14 A	17. Feb.	5.Mar.	20. A	29.Maij
3693	8	*		d	3.14.A	3 A	1. Feb.	18.Feb.	15 A	14.Maij
3694	9	xxj		c	23.22.M	23 M	24. Ian.	10.Feb.	28. M	6.Maij
3695	10	ij		b	11.20.A	11. A	13.Feb.	1.Mar.	17. A	26.Maij
3696	11	xiij	A	g	31. 5.M	31. M	29. Ian.	15.Feb.	1. A	10.Maij
3697	12	xxiiij		f	19. 2.A	18. A	17.Feb.	6.Mar.	21. A	30.Maij
3698	13	v		e	8.11.A	8 A	9. Feb.	26.Feb.	13. A	22.Maij
3699	14	xvj		d	28.20.M	28. M	25. Ian.	11.Feb.	29. M	7.Maij
3700	15	xxvj		c	17.17.A	17. A	14. Feb.	3 Mar.	18. A	27.Maij
3701	16	vij		b	6. 2.A	6. A	6.Feb.	23.Feb.	10. A	19.Maij
3702	17	xviij		A	26.11.M	26 M	29. Ian.	15.Feb.	2. A	11.Maij
3703	18	xxix		g	14. 9.A	14 A	11.Feb.	28.Feb.	15. A	24.Maij
3704	19	x	f	e	2. 7.A	3. A	3 Feb.	20.Feb.	6. A	15.Maij
3705	1	xxij		d	23. 2.M	22. M	25. Ian.	11.Feb.	29. M	7.Maij
3706	2	iij		c	11. 0.A	10. A	7.Feb.	24.Feb.	11. A	20.Maij
3707	3	xiiij		b	31. 9.M	30. M	30. Ian.	16.Feb.	3. A	12.Maij
3708	4	xxv	A	g	18. 6.A	18. A	19. Feb.	7.Mar.	22. A	31.Maij
3709	5	vj		f	7.15.A	7 A	10. Feb.	27.Feb.	14. A	23.Maij
3710	6	xvij		e	28. 0.M	27. M	26. Ian.	12.Feb.	30. M	8.Maij
3711	7	xxviij		d	15.21.A	15. A	15. Feb.	4 Mar.	19. A	28.Maij

Anni

14. Page of the Gregorian Easter table

14 Nisan but the 15th and in the Diaspora the 16th; in consequence, many Christians imagined they were forbidden to celebrate Easter on those dates (though Luther had taken the Passover of 16–17 March 1538 as confirmation of his case). Whereas 15 Nisan, at least according to the current Jewish calendar, had not coincided with Julian Easter since 783, the 16th not since 1315, Gregorian Easter, despite the displaced epacts, fell on 15 Nisan in 1609 and on the 16th six times before the end of the 16th century. Protestants did not fail to complain.

Astronomical Easter

In 1699 the German Lutherans voted to accept the New Style, by passing directly from 18 February 1700 to 1 March, but not the Gregorian Easter tables. Instead, they decreed that Easter should be governed by the real spring equinox and the real full moon as indicated in the best astronomical tables available, based on the meridian of Tycho Brahe's laboratory at Uraniborg in Denmark. This method was known as the *calculus astronomicus*. (John Dee in 1582 had already suggested that astronomically based Easter tables should be drawn up for the meridian of London.) The reform was adopted by Denmark, most of the Protestant Swiss cantons, and, under the name 'Improved Julian Calendar', by the remaining Dutch provinces.

Difference usually arose when the real full moon fell on a Saturday, but the Gregorian tables indicated the following Sunday, entailing a week's postponement of Easter. This happened in 1700 itself, when only the Gregorian date was observed; but the next discrepancy, in 1724, was the subject of anxious debate. The Pope's tables set down *luna XIV* on 9 April, again a Sunday, entailing Easter on the 16th, but once more the real new moon fell the day before. Easter would therefore, according to the *calculus astronomicus*, be the 9th, one week earlier than in Roman Catholic countries.

However, a difficulty was raised that was none the less keenly felt for being spurious, namely coincidence with the second day of Passover. That led to a re-examination of the supposed prohibition; the outcome was that German and Swiss Protestants kept the astronomical date, though Denmark did not. (The Dutch Protestants, in this and all other cases, kept the Gregorian date.) As a result, Bach's *St John Passion* was first performed in Leipzig on astronomical Good Friday, 7 April 1724, a week ahead of the Gregorian Good Friday – and also, as it happened, of Julian Good Friday on 3 April Old Style.

Sweden had held aloof in a failed attempt to adopt the New Style painlessly; in 1740 the *calculus astronomicus* was introduced, but Old Style dating remained till 1753. Thus in 1742, when Protestant and Roman Catholic Easter both fell on 25 March New Style, in Sweden it was called the 14th. Two years later, another discrepancy arose between astronomical and Gregorian Easter; this time not only German and Swiss Protestants but Denmark too kept the astronomical date, 29 March. Sweden kept the same day, but called it the 18th.

Before the next discrepancy, in 1778, the German Protestants, followed by the Swiss and the Danes, abandoned the *calculus astronomicus* at the behest of Frederick the Great, who had acquired a large number of Roman Catholic subjects through the First Partition of Poland (1772); his insincere pretext, coincidence with Passover, also deterred the Swedes from observing the astronomical date in 1778 and 1798, but they did keep it three times in the early 19th century, before abandoning it in 1823. Finland, part of Sweden till 1809 and then ceded to Russia, did so a further three times, the last in 1845.

The case of 1798, however, illustrates a difficulty inherent in the astronomical Easter: although at Uraniborg full moon occurred just before midnight on Saturday, 31 March, in most of Sweden it was already Sunday, 1 April, so that observance of the astronomical Easter on that day would have entailed celebrating on the day of full moon, a major impropriety. This case must occasionally arise in the *calculus astronomicus* whatever the meridian chosen.

Great Britain

When Great Britain adopted the New Style, the political necessity of crediting the reform to Parliament and not the Pope required that the Church of England should be spared the humiliation of accepting the Gregorian Easter tables; on the other hand, High Churchmen would have objected to adopting the Lutheran astronomical Easter used in George II's electorate of Hanover. It was therefore necessary to achieve the Pope's results by different means.

The solution was to devise tables (to be found in the Book of Common Prayer) for reassigning the Golden Numbers to different dates each century. This plan had been considered by the papal reformers, but rejected because, in order to show the lune of every day in the year, thirty different tables would have been needed; that rejection made it all the more attractive to the Church of England, which had no interest in the lunar calendar except for finding Easter. For the same reason, the Golden Numbers, which previously had been written against the year's new moons, were now made to mark directly the Paschal term or 'Ecclesiastical Full Moon'. The Church of England thus always celebrates Easter on the same day as the Church of Rome, but without breathing a word about epacts.

The Orthodox churches

The most sensitive proposal in the 'Revised Julian Calendar' that certain Orthodox Churches approved in May 1923 was the upset to the centuries-old tradition of the 'Nicene' – or rather Alexandrian – Easter, the feast of feasts, by adopting the *calculus astronomicus* based on the meridian of Jerusalem. Although for a few years after 1923 some churches kept non-Julian dates, tradition soon reasserted itself; Easter is still kept according to the Julian calendar by almost all Orthodox churches (the Finnish minority church – though not that of the Russian community – and a few Western parishes have adopted the Gregorian calendar in full).

In 1997 another decision was taken to adopt the astronomical Easter from 2001, when it would coincide with the Julian as well as the Gregorian date; but in 2002 it was the Julian date that was kept. If no further reform takes place in either calendar, from 6700 to 6799 Orthodox Easter will coincide with Western Pentecost, though the nominal dates in the Revised Julian Calendar will be one day later.

Fixed Easter

Certain early Christian communities had kept a fixed Easter on 25 March or 6 April, respectively the 14th of the Cappadocian month Teireix and of the 'Province of Asia' month Artemision; the former had the advantage of being not only the feast of the Annunciation but the traditional date of the Crucifixion in the Western church and the Resurrection in the Eastern. Luther recommended that Easter should be made immovable like Christmas; but the notion of Easter on a day other than Sunday has had no subsequent appeal.

In 1723, observing that in the next year Protestants would keep a different Easter from Roman Catholics, the Swiss mathematician Jean Bernoulli proposed that the feast should always be the first Sunday after 21 March; in 1834 Marco Mastrofini, after setting forth his scheme for an invariable calendar (see Chapter 5), suggested more tentatively that if it were accepted, Easter should be fixed as Sunday, 2 April. In 1926 the League of Nations recommended that the feast should be kept on the Sunday after the second Saturday in April; in the United Kingdom this provision was incorporated in an Act of Parliament, to take effect upon general agreement amongst the churches. Such agreement has not been reached.

Chapter 5
Weeks and seasons

As we saw in Chapter 3, the ancient Romans had an eight-day market-cycle, the *nundinum*, independent of months and years, registered in their calendars by labelling each day with a letter from A to H. This acquired an ultimately victorious rival in the seven-day cycle known as the week. However, the week as we know it is the fusion of two conceptually different cycles: the planetary week, originally beginning on Saturday, derived from Hellenistic astrology, and the Judaeo-Christian week, properly beginning on Sunday.

Whereas astrology was alien to both Greek and Egyptian tradition, in Babylonia planetary observations had long been used to predict affairs of state; during the 5th century BC the principles were extended to predict the fates of individuals. By then, both Babylonia and Egypt were part of the Persian Empire; although Egypt for a time regained independence, it was reconquered shortly before Alexander the Great's defeat of Persia brought about the cultural and political upheaval across the known world that enabled a would-be science of the future to spread, and with it the principle of planetary dominion.

It was from this time on that astrologers, first in Egypt and then elsewhere, held every hour to be under the domination of a planet according to the inward sequence from Saturn to Moon; furthermore each day was governed by the planet of its first hour

15. Saturn and his day. The hours of Saturday (beginning on the right with the daytime hours) are marked *N(ocens)*, 'harmful', if under Saturn or Mars; *b(ona)*, 'favourable', if under Jupiter or Venus; *c(ommunis)*, 'neutral', if under Sun, Mercury, or Moon. The text underneath reads: 'When it is Saturn's day or his hour by night or day, all things become dark and difficult; those who are born will be in danger; he who disappears will not be found; he who takes to his sickbed will be in danger; stolen goods will not be recovered.'

(see Figure 15). Since the 24 hours of the natural day accommodated three planetary cycles with three hours, and therefore three planets, left over, the next day was ruled by the next planet but two: after Saturn the Sun, after the Sun the Moon, and so on (see box).

This planetary week would spread east to India and China, and west to Rome, where it attained written record in the reign of Augustus (sole ruler 31 BC–AD 14), under whom the poet Tibullus mentions 'the day sacred to Saturn' and an inscription presents the eight-letter A–H cycle of the *nundinum* accompanied by a seven-letter A–G cycle for the week (see Figure 16).

The origin of the planetary week

♄ = Saturn, ♃ = Jupiter, ♂ = Mars, ☉ = Sun, ♀ = Venus, ☿ = Mercury, ☽ = Moon, D = day, N = night

Bold marks the presiding planet of the day
Hour Presiding planet

	D	N	D	N	D	N	D	N	D	N	D	N	D	N
1	**♄**	☿	**☉**	♃	**☽**	♀	**♂**	♄	**☿**	☉	**♃**	☽	**♀**	♂
2	♃	☽	♀	♂	♄	☿	☉	♃	☽	♀	♂	♄	☿	☉
3	♂	♄	☿	☉	♃	☽	♀	♂	♄	☿	☉	♃	☽	♀
4	☉	♃	☽	♀	♂	♄	☿	☉	♃	☽	♀	♂	♄	☿
5	♀	♂	♄	☿	☉	♃	☽	♀	♂	♄	☿	☉	♃	☽
6	☿	☉	♃	☽	♀	♂	♄	☿	☉	♃	☽	♀	♂	♄
7	☽	♀	♂	♄	☿	☉	♃	☽	♀	♂	♄	☿	☉	♃
8	♄	☿	☉	♃	☽	♀	♂	♄	☿	☉	♃	☽	♀	♂
9	♃	☽	♀	♂	♄	☿	☉	♃	☽	♀	♂	♄	☿	☉
10	♂	♄	☿	☉	♃	☽	♀	♂	♄	☿	☉	♃	☽	♀
11	☉	♃	☽	♀	♂	♄	☿	☉	♃	☽	♀	♂	♄	☿
12	♀	♂	♄	☿	☉	♃	☽	♀	♂	♄	☿	☉	♃	☽

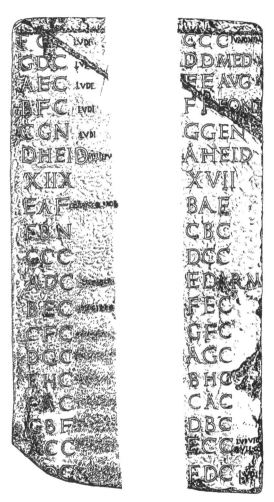

16. Fragments of Fasti Sabini showing weekday letters beside nundinal letters

An early stage in the victory of week over *nundinum* is presented by a graffito at Pompeii giving the market-days in various towns and cities: although eight place-names are listed, beginning with Pompeii and ending with Capua (Rome is seventh), the neighbouring column presents the days of the week from 'Sat.' to 'Ven.' Unfortunately the writer wrote column by column rather than line by line, so generously spacing the days and so tightly cramping the cities that we cannot even be sure whether he noticed the discrepancy (Figure 17).

Jews, meanwhile, had been observing a week of their own, in which six working days, numbered from 1 to 6, were followed by a rest day, or *shabbat*, in English 'Sabbath'. *Shabbat* coincided with the astrological Saturday, which was a day of ill-omen because Saturn was a baleful planet; for this reason, pagan writers misrepresent it as a joyless fast-day. To be sure, Sabbath regulations, in some quarters at least, were sterner than they have since become; the Book of Jubilees, for instance, forbids married couples to make love on that day, which the rabbinical tradition positively encourages them to do, but even this text prohibits fasting. Nevertheless, these pagan accounts seem more appropriate to the Babylonian day of ill-omen on the 7th, 14th, 21st, and 28th of the month, in which some writers have seen the origin of the Sabbath, but which would seem rather to have been adapted to the astrological Saturday, subjected to the most malign of planets.

Shabbat was the last day of the Jewish week, but corresponded to the first day of the planetary week. However, since nobody likes an inauspicious beginning, by the 2nd century AD the astrologer Vettius Valens was reckoning the planetary week from Sunday. Although the Sun was not an auspicious planet like Jupiter or Venus, neither was it malign, like Saturn or Mars (see box, p. 70). The change also had the effect of aligning the planetary and Jewish weeks; in a society that used the Jewish divine name Iao in its magic, this may not have been accidental.

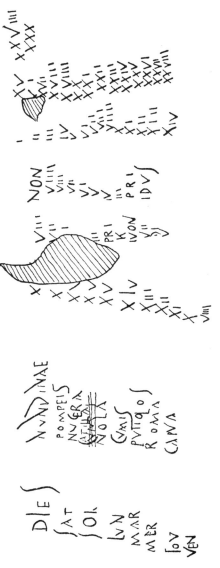

17. Graffito from Pompeii showing conflation of week and market-cycle; the right-hand half presents days of two months and lunar ages

By then Christians had adopted the Jewish week with its numbered days, though for 'first day' they said 'Lord's Day'. For Greek-speakers Monday to Thursday were the second to fifth day respectively and Friday the Preparation (from John 19: 31), Latin-speakers called Monday to Friday the second to sixth *feria* respectively, extracting from classical *feriae* 'holiday' a new singular that is still used in technical discussions of the calendar in preference to the circumlocution 'day of the week'. The seventh day remained 'Sabbath' in both languages, but the Jewish rules were not observed by the Gentile church; at Rome it even became the custom to fast on that day, which shocked the Christians of the East.

Evidence from Roman Egypt shows courts observing a weekly break on Thursday, enjoying the benevolence of its presiding planet Jupiter; but in 321 the Christian emperor Constantine decreed that suits should not be heard on 'the day celebrated for the veneration of its Sun'. He did not use the Christian term 'Lord's Day' (the majority of the Empire's citizens were still pagan), though a court record from four years later does so; later laws sometimes combine the two names. These ordinances converted Sunday from a weekly remembrance of the Resurrection to a day of rest, though even in the 6th century Bishop Caesarius of Arles was complaining that the local peasantry rested on Thursday instead. (This was till recently the traditional day off in French schools.)

In the Christianized empire the planetary names competed with the Christian, with varying outcomes to be discerned from the modern languages (see box). In Greek and Portuguese the Christian terminology triumphed; in other Romance languages Sunday and Saturday became 'Lord's Day' and 'Sabbath', but the other planetary names in general refused to be dislodged (though in Sardinia Friday became *kenapura*, 'pure supper', commemorating the Last Supper). The Britons proved even more resistant to the Christian terms; all seven planetary names are easily recognized in Welsh, Cornish, and Breton.

Names of weekdays

Pagan Greek	Church Greek	Pagan Latin	Church Latin	French
Hēlíou*	Kyriakḗ	dies Solis†	dominicus/a	dimanche
Selḗnēs	Deutéra	dies Lunae	secunda feria	lundi
Áreōs	Trítē	dies Martis	tertia feria	mardi
Hermoû	Tetártē	dies Mercurii	quarta feria	mercredi
Diós	Pémptē	dies Iovis	quinta feria	jeudi
Aphrodítēs	Paraskeuḗ	dies Veneris	sexta feria	vendredi
Krónou	Sábbaton	dies Saturni	sabbatum	samedi

* Understand *hēméra* 'day' with name. † Or *Solis dies*, and so throughout.

Portuguese	Welsh	English	German	Polish
domingo	dydd Sul	Sunday	Sonntag	niedziela
segunda-feira	dydd Llun	Monday	Montag	poniedziałek
terça-feira	dydd Mawrth	Tuesday	Dienstag	wtórek
quarta-feira	dydd Mercher	Wednesday	Mittwoch	środa
quinta-feira	dydd Iau	Thursday	Donnerstag	czwartek
sexta-feira	dydd Gwener	Friday	Freitag	piątek
sábado	dydd Sadurn	Saturday	Samstag	sobota

The Germanic peoples stood almost as firm: as the Romans had substituted Roman deities for Greek, so they substituted their own: for Mars, either the warrior god called Tīw in English and Týr in Norse, or a god of the warband in assembly (*thing*) found on inscriptions as Mars Thincsus (whence Dutch *Dinsdag*, German *Dienstag*); for Mercury, as god of wisdom, Odin; for Jupiter, as thunder-god, Thor; for Venus, the love-goddess Frigg. In English and Dutch no further change has been made; Saturn, for whom the Germanic pantheon had no equivalent, is left in place as lord of the seventh day. In German there are modifications: Wednesday is 'midweek' (*Mittwoch*) as being halfway between Sunday and Saturday, Saturday either the sabbath-day (*Samstag*) or Sun(day)-eve (*Sonnabend*). The former, regularly used by Roman Catholics, is now displacing the latter amongst Protestants, being a more natural expansion of the abbreviation *Sa*. In dialects other forms are found; Bavarian has *Ertag* for Tuesday, from pagan Greek *Áreōs*, 'of Ares', and *Pfinztag* for Thursday, from Christian Greek *Pémptē* 'fifth'.

The Slavs adopted a different system. Sunday was called 'not-work', *nedělja*; this underlies the name used in all modern Slavonic except Russian, in which it means 'week' and Sunday is *voskresen'e*, 'resurrection' (a more learned form of the word, *Voskresenie*, means 'Easter'), though Monday is still 'the one after *nedělja*', *ponedel'nik*. Saturday became 'Sabbath', but though Wednesday is 'middle' (compare *Mittwoch*), Tuesday, Thursday, and Friday have names derived from 'two', 'four', and 'five' respectively, counting not from Sunday as in Church Greek and Latin but from Monday, which is the first day of the Orthodox liturgical week except between Easter and Pentecost. Lithuanians (mostly Roman Catholics) and Latvians (mostly Lutherans) also count from Monday, the first day of the week for commercial and administrative secularists; by contrast, the Society of Friends used to number the days of the week from first to seventh from Sunday to Saturday.

Islam adapted the Jewish week by making Friday the day of prayer (not of rest, except in imitation of the West) with the Arabic name of *jum'a*, 'gathering'; the Sabbath remains in name though not in nature, and the other feriae are still numbered from Sunday. (In Kiswahili Saturday to Wednesday are days 1–5, while Thursday is Alhamisi, Arabic for 'the fifth'.)

Sunday Letters

The A–G cycle written in the Middle Ages and long afterwards against the days of the year in calendars and almanacs made finding the feria of any date easy; if today was Tuesday, one looked at its letter in the calendar and knew that other days with the same letter would also be Tuesday until either the *bissextus* (which shared the letter F with the regular *VI K. Mart.*) or the New Year intervened. It was made even easier by mnemonics that gave the letter for the 1st of every month; most were in execrable Latin, but an English specimen runs:

> At Dover Dwells George Brown, Esquire,
> Good Christopher Fitch And David Frier.

That is, 1 January is A, 1 February D, and so on; hence any quantième in February will be three letters later in the cycle than the same quantième in January.

Churchmen, interested in identifying Sundays, and especially Easter Sunday, called these letters *litterae dominicales* or Sunday Letters. The same term is used for the letter corresponding in any given year to Sunday (see box); Easter is the first Sunday after *luna XIV* to bear the Sunday Letter of the year, between 22 March (D) and 25 April (C).

Assaults repelled

The French Republican calendar (see box) had no place for the Christian week; instead, in accordance with the decimal principle beloved of the revolutionaries, months were divided into three 10-day *décades*, each ending with a *fête décadaire* dedicated to an entity ranging from the French People to Misfortune.

The competition between the two calendars was represented in pamphlets as a conflict between the revolutionary Citoyen Décadi and the reactionary Monsieur Dimanche. In 1798 the new calendar was made compulsory, all observance of the week being forbidden. Public workers who took Sunday off were to be dismissed; it was even attempted to impose *décadi* on the Church instead of Sunday as the day of worship. In much of France the campaign against the week was a failure from the outset; it was given up by Napoleon even while the *calendrier français* remained officially in force.

The French Republican calendar

In 1793 the Convention enacted a new calendar on Alexandrian lines, backdated to the establishment of the French Republic on 22 September 1792. The year comprised 12 months of 30 days plus 5 *jours complémentaires*, or 6 before the Julian leap year; hence year VII (1798/9) was a leap year (*année sextile*) even though Gregorian 1800 was not. The month names reflected the French climate and the French agricultural year, with a different suffix for each season: autumn *-aire*, winter *-ôse*, spring *-al*, and summer *-dor*.

Another challenge to the week came from the Soviet Union. In the new calendar of 1929, which gave all 12 months the same length of 30 days, interspersed with five national holidays, the seven-day week was replaced with a five-day cycle from Monday to Friday (or 'first' to 'fifth') without the 'bourgeois idlers' Saturday and Sunday; each of the five days was a rest-day for one-fifth of the population, every citizen receiving a coloured slip corresponding to his or her day off. (The national holidays were excluded from the cycle.)

The scheme was intended to combine generous individual leave with continuous production, in which it failed, and to disrupt religious observance and family life, in which it succeeded well enough to provoke excessive discontent. As a result, the reform was modified: from 1 December 1931 the traditional months were restored, and also the uniform rest-day, but not the week; instead the rest days were to be the 6th, the 12th, the 18th, the 24th, and (except in February) the 30th of each month. Although the regime persuaded Westerners sympathetic to communism or to calendar reform that urban workers no longer remembered the seven-day week, the peasantry sabotaged it by taking both Sundays and the new rest-days off; in 1940 Stalin restored the seven-day week with Sunday rest.

Besides these revolutionary schemes, reformist proposals have been made from time to time for standardizing the relation between date and feria by excluding one day from the week in each year, or two days in leap year; such days are termed *blank days*. Although in 1834 Marco Mastrofini merely proposed to exclude 31 December and leap day (which was to follow it) from the week, so that 1 January should always be a Sunday, and then to fix Easter on 9 April, other reformers envisaged broader schemes to regularize the months or at least the seasons.

As early as 1745, in a letter to the *Gentleman's Magazine*, 'Hirossa Ap-Iccim' proposed that the year should contain 13 months, each comprising four weeks, followed by a blank day for Christmas and a national day in leap year; that 11 December 1745 Old Style should become 1 January 1750 in a new era beginning in 4 BC; that leap year should be suppressed every 132 years; and that weights, measures, and money should be reckoned in multiples of eight. The author was a Maryland clergyman, the Revd Hugh Jones, who dedicated a fuller exposition of these notions to the Earl of Chesterfield (British Library, Add. MS 21893) and published them over the name 'H. J.' as *The Pancronometer* (London, 1753). No notice was taken; however, a 13-month year was the basis of the Positivist calendar, with epoch 1789, promulgated by Auguste Comte in 1849 and still enjoying some currency in France and Brazil. Months and days were dedicated to great men and the occasional great woman; the 365th day is dedicated to the dead, leap day to holy women.

In the early 20th century the 13-month year was adopted for internal accounting by several companies. On this basis two businessmen in the United States, the English-born Moses B. Cotsworth and George Eastman of Eastman Kodak, set about converting the world to an International Fixed Calendar, better known as the Eastman Plan, in which a 13th month called Sol should precede July, and blank days should be observed on 29 December (Year Day) and in leap year 29 June (Leap Day). Far too hardheaded to shrink from a year with 13 Friday the 13ths, Eastman observed that 13 was the lucky number of a country founded by 13 insurgent colonies.

Outside the United States, this argument naturally cut less ice. Some reformers toned down the reform by proposing 12-month years in which every third month should comprise five weeks instead of four; but more attention was paid to a 'World Calendar' in which March, June, September, and December should have 31 days and the rest 30, again with blank days after December and in leap year after June.

In the 1920s the League of Nations took an interest in both the Eastman Plan and the World Calendar; but the interruption to the week proved a stumbling-block. In 1931 the Chief Rabbi of the United Hebrew Congregations of the British Empire, Dr Joseph Hertz (see Figure 18), delivered an impassioned attack on the reforms at League headquarters in Geneva, emphasizing the impossibility of Orthodox Jews' accepting an eight-day interval between Sabbaths, and the inconvenience they would suffer by observing the Sabbath on another day than Saturday. His intervention had an effect best gauged by the bitter resentment it aroused among secularist reformers.

Seventh-Day Adventists also objected to the disruption, and the imperial government of India, aware of the many sensibilities that would be offended, declared it unacceptable. By contrast, when the World Calendar was briefly revived in the 1950s, it was favoured by the committed secularists who ruled that country after independence; however, no major power saw any advantage in the reform, and since then campaigners for a better world have directed their energies to more pressing concerns. The week thus remains

18. Portrait of Dr Joseph Herman Hertz (1872—1946), the Chief Rabbi
who fought to save the week

the oldest calendrical institution still to function without structural change.

The week-based year

When Dr Hertz addressed the League of Nations, he declared that Jews had no objection to calendar reform as such, provided that it left the week untouched; in particular, if a fixed relation between feria and date were intended, it could be achieved by a year of 52 weeks or 364 days, with a leap week added from time to time. This met with no approval, even though a 364-day year with leap weeks had been known for centuries in Iceland, where it regulated civil affairs; it comprised 12 months of 30 days (or rather nights), four epagomenal days or ekenights (*aukanætr*), and in five years out of 28 an ekeweek (*aukavika*). It was based on the old Germanic two-season alternation of 'summer' and 'winter', but was correlated with the Church calendar, from which it adopted the week and the solar cycle; indeed, dates were normally reckoned not by months (except in the latter part of winter, the very portion of the year King Romulus had supposedly not provided months for) but by the number of weeks remaining before or elapsed after midsummer or before midwinter.

Dr Hertz, however, may have had in mind the 364-day year used or advocated by some Jews during the Second Temple Period comprising 12 months, of which every third month contained 31 days and the rest 30; the week began not on First Day (Sunday) but on Fourth Day (Wednesday), the day on which the sun and moon were created. It is disputed how ancient this calendar was (some scholars even suggest it, or something like it, was the pre-Exilic norm), nor how widespread its use was at any time; it is also disputed whether ekeweeks were added, or adherents took the view, if calendar and sun were out of step, so much the worse for the sun.

Other groupings

Old Irish legal texts recognize groups of 5 and 15 days, the latter surviving from the half-months of the pre-Christian lunar calendar. Market-cycles 3, 4, 5, 6, 9, or 10 days long are or have been used in many parts of the world, besides the Roman 8-day *nundinum*; more than one people of Nigeria makes use of the 4-day cycle and its 8- and 16-day multiples in several contexts.

When towns or villages hold markets on different days of the same cycle, this constitutes a link between them; on the other hand, when as notably in West Africa more than one market-cycle coexists in a particular locality, there may be especial solidarity between those villages that share a cycle and rivalry between those that do not, particularly if they hold markets on the same day.

The week, and the analogues discussed so far, run concurrently with months and years; Sunday the 1st is followed by Monday the 2nd, Wednesday 31 December 2003 by Thursday 1 January 2004. In some cultures, however, the quasi-weeks themselves run concurrently: in Central America, before the Spanish conquest, the cycles of 13 and 20 days that formed the 260-day cycle did so (see Chapter 6), and no fewer than 9 concurrent cycles, from 2 to 10 days long, are found in Indonesia; of these the 5-, 6-, and 7-day cycles are the most important, combining to form a 210-day *odalan*.

Seasons

The Western world has inherited from the ancient Romans the division of the year into four seasons or 'times of year' (*tempora anni*, whence German *Jahreszeiten*): few Latin words are easier to translate into modern European languages than *ver aestas autumnus hiems*. Diagrams of the cosmos show these seasons integrated into a world-model (see Figure 19); they reappear in another temperate-zone culture, that of China, as *chūn xià qiū dōng*.

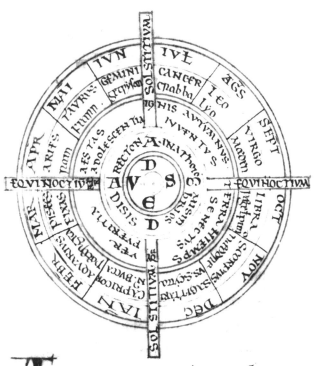

 Æften iuniuṁ cymð iuliuṛ be
hæpð an ⁊ þʀitig ðaga æften þæne
sunnan ʀyne . ⁊ æften þæṛ monan
þʀitig . ⁊ xv. lêt augṛ . gæð seo sunne
on þ tacen þeyṛ genemneð leo .

19. Diagram from Byrhtferth's *Enchiridion* (early 11th century),
mapping cardinal points, seasons, signs of the zodiac, and ages of man
onto months

Yet this division is by no means universal. In Nigeria the Yoruba speak of two half-year seasons, dry and rainy; but India has six seasons, whose Sanskrit names are *grīṣma* ('hot season'), *varṣa* ('rainy season, monsoon'), *śarad* ('autumn'), *hemanta* ('winter'), *śiśir* ('cool season'), *vasanta* ('spring'). Each comprises two months; these differ across the country, but in the National Calendar (see Chapter 6) spring begins in Phalguna.

Ancient Egypt recognized three seasons: flood, winter, and summer, each comprising four months of the solar year, which until the 6th century BC were simply numbered as the so-manyeth month of such-and-such season (see Appendix A). However, the failure to intercalate meant that for much of Egyptian history nominal and real seasons did not correspond. In 824 BC, for instance, when the New Year began on 21 March, 'winter' ran from the theoretical beginning of flood on 19 July to 15 November, followed by 'summer'.

Classical Greek knows spring (*éar*, or in a dialectal variant *wêr*), summer (*théros*), and winter (*cheimṓn*), but a distinct concept of and name for autumn (*phthinópōron*, 'waning harvest') took a long time to emerge. However, the great Greek historian Thucydides (5th–4th century BC), though familiar with spring, divided the years of the Peloponnesian War into summer (the campaigning season) and winter.

A similar two-season conception existed amongst the Germanic peoples, proving especially resilient not only in Iceland, but also in Scandinavia, where summer is said to begin on St Tiburtius' day (14 April) and winter on St Callistus' day (14 October). In consequence, although the Germanic languages all share the words 'summer' and 'winter', they had no words for 'spring' and 'autumn' until they encountered Roman culture. The names for those two seasons vary considerably between languages, and even within them: whereas British English has retained the Latin or French 'autumn' used by Chaucer, Tyndale, and Shakespeare, the US uses 'fall', short for 'fall of the leaf', first attested in the 16th century. In German, spring may be either *der Frühling* or *das Frühjahr*; an archaic or poetic term *Lenz* corresponds to current Dutch *lente* and Old English *lencten*. This latter was used both of the season and of an ecclesiastical fast; the fast is still called by the shortened form *Lent*, of which *Lenten* (as in *Lenten sermon*) is now understood as the adjective; the season has since the 16th century been 'spring', short for 'spring of the year'.

In the Celtic languages too 'spring' and 'autumn' vary, but there are shared words for 'summer' and 'winter'; the two-season division is explicitly stated in an Irish glossary. Irish *samhradh* and Welsh *haf* are related to English *summer*, *geimreadh* and *gaeaf* to Latin *hiems*; the same words underlie the Gaulish month names Samonios and Giamonios (see Chapter 6).

Latin writers offer various dates for the beginnings of the seasons: the Elder Pliny (d. AD 79) gives 8 February, 10 May, 11 August, 11 November; Isidore of Seville (d. 636) gives 22 February, 24 May, 23 August, 23 November; the Venerable Bede, writing in 725, gives 7 February, 9 May, 7 August, 7 November, in other words, the seventh day before the Ides of the respective months. Both Isidore's and Bede's dates are commonly given in medieval calendars; on Bede's scheme, the equinoxes and solstices fall about the middle of their seasons, so that the Nativity of the Baptist is Midsummer's Day. Likewise, the normal Old English name for Christmas was 'midwinter'.

In Ireland spring is reckoned from 1 February, summer from 1 May, autumn from 1 August, and winter from 1 November; of these dates 1 February is St Bride's, or Brigid's, day, to the obliteration of the pre-Christian *Imbolc* (in modern Irish *Oímelc*) that originally marked the new season, but the other three retain their ancient names of *Bealtaine*, *Lúnasa* (formerly *Lughnasadh*), and *Samhain*. The *Oxford English Dictionary* regards this as the 'British' disposition, as opposed to a 'North American' practice of beginning the seasons with March, June, September, and December; but most Britons would prefer the latter, which is the scheme adopted by the Meteorological Office.

Naturally, no system based on natural phenomena can be valid all over the globe, or even within the northern hemisphere; more objective is the principle of beginning seasons with the equinoxes and solstices, hence currently on or about 20 March, 21 June, 22 September, and 21 December. These are often described as the 'official' beginnings of the respective seasons, although no royal proclamation or Act of Parliament has so decreed. Neatness of scheme often trumps truth to fact: the 11th-century diagram shown above in Figure 19 blithely equates seasons determined on this principle with three-month blocks beginning in April, July, October, and January.

From the time of Caesar's reform, Roman tradition equated these cardinal points with the eighth day before the Kalends, on 25 March, 24 June, 24 September, and 25 December, dates that in fact had applied some two centuries earlier. The Church incorporated them into its calendar as respectively the Annunciation (the Conception of Jesus Christ), the Nativity of St John the Baptist, his Conception (shifted in the East to the 23rd; see Chapter 6), and the Nativity of Jesus Christ. However, the Western church preferred to begin the seasons with fasts on the Wednesday after Quadragesima, Whit Sunday, Holy Cross day (14 September), and St Lucy's day (13 December), each resumed on the following Friday and Saturday. These are known in English as the Ember fasts, a corruption (compare German *Quatember*) of Latin *quattuor tempora*, 'the four seasons'. (The Church of England has changed the weeks.)

Although in the southern hemisphere summer corresponds to northern winter and vice versa, European settlement has imported many northern associations of date and season. Papai Noel is cruelly sent out into the heat of a Brazilian summer wearing his heavy red suit and white fur; but Latvians in Australia celebrating St John's day in the winter cold of 24 June are glad of their wool-based national costume.

Chapter 6
Other calendars

The Jewish calendar

The modern Jewish lunar calendar, based on calculation like the Gregorian but far more complex, evolved out of one used after the return from Babylonian exile, based on observation. Before then the months, though occasionally given Phoenician names, were normally numbered; the count began in spring even though the harvest is said in Exodus to come at the end of the year. It is disputed whether the calendar was lunar or solar, and in the latter case what relation it bore to the post-Exilic 52-week year discussed in Chapter 5.

After the Exile, the old habit of numbering months gradually gave way to use of the Babylonian names (but also of Macedonian names when Jews were speaking Greek); the spring New Year, on 1 Nisan, and the autumn New Year, on 1 Tishri, continued to compete for many centuries. The latter was always used to reckon the seven-year sabbatical cycle, in the final year of which cultivation was not permitted; the rule of thumb for biblical exegesis was that for 'kings of Israel' years were counted from Nisan, for other kings from Tishri, but exceptions could not be denied. The eventual compromise was that 'Nisan is the head of the months, Tishri is the head of the year': although the months specified in the Torah for the major festivals are counted from the spring, the definitive civil year begins in the autumn.

So long as the new moon was determined by observation, if on the 30th night the crescent was reliably reported and confirmed by the Sanhedrin, then the day (beginning at sunset) was reckoned as the first of the month, the previous month being hollow. Otherwise it was treated as the 30th of the old month, and the new month began the following night. A certain degree of manipulation was admitted to prevent fasts, in particular Yom Kippur, from falling on the day before or after the Sabbath (there was no objection to coincidence with the Sabbath itself); as a result 1 Tishri could not be a Wednesday or a Friday. Considerably later it became a rule that 1 Tishri must not fall on a Sunday either, since the ritual for the last day of Succoth (Tabernacles), on 21 Tishri, was deemed too energetic for the Sabbath. There was also a rule that no year should have fewer than four full months or more than eight; in embolismic years this limit was raised to nine.

Embolism consisted in repeating the last month, Adar, of the spring-to-spring year, when it appeared that otherwise Passover would come too early. Before the Roman destruction of the Temple in AD 70 there seems to have been a preference for late Passovers, so that the pilgrims should have plenty of time to reach Jerusalem; afterwards, when early Passovers no longer caused difficulty, it became the rule to intercalate when any two of three conditions should obtain: that the crops were still young, that the fruit-trees were not ripe, that the festival would otherwise precede the equinox. There was also a preference for intercalating pre-sabbatical years rather than sabbatical or post-sabbatical ones.

Contrary to Christian suppositions, it was not an absolute rule that Passover should follow the equinox, though it must usually have done so while the Temple stood. The assertion that Jews were not observing their own rules had more justification in the readiness of Diaspora communities to regulate the calendar for themselves, rather than following the rabbis of Jerusalem: at Antioch, for instance, the rule was that 14 Nisan should fall within the civil month of Dystros, which was the local name of Roman March.

Tradition, unsupported by evidence, has it that in AD 359 observation was replaced by calculation because the Romans were obstructing the messengers sent from Jerusalem to announce new moons and intercalations. In fact, however, a variety of practices, including observation and calculation not in accordance with the modern rules, persisted long afterwards; not till the 10th century was full uniformity achieved.

The modern calendar operates in two stages: first, the *molad* or 'birth' – the first conjunction – of autumn is found, then the date of 1 Tishri derived from it. The day, subdivided into 24 equal hours each comprising 1080 'minims' of 76 'moments', is reckoned from sunset, which for calendrical purposes – but not for civil or religious life – is normalized as 6 p.m. Jerusalem time (3.39 p.m. at Greenwich); the synodic month is defined as 29 days 12 hours 793 minims. The common year of 12 months thus contains 354 days 8 hours 876 minims; the embolismic year has 13 months, making 383 days 21 hours 589 minims.

Embolisms take place in years 3, 6, 8, 11, 14, 17, and 19 of a Metonic cycle, comprising 235 lunations = 6939 days 16 hours 595 minims; the cycles are reckoned from the Creation at 5 hours 204 minims on Monday, 7 October 3761 BC, which in our midnight-to-midnight reckoning would be 11 hours 11 minutes 20 seconds p.m. on Sunday the 6th. The *molad* of year Y will thus take place as many days, hours, and minims after that epoch as are contained in the $(Y-1)$ theoretical years elapsed.

Once the *molad* has been found, a date for 1 Tishri is sought at a permitted distance from the previous 1 Tishri and the next. There are six possible lengths of the year:

common deficient	353 days	embolismic deficient	383 days
common regular	354 days	abundant regular	384 days
common abundant	355 days	embolismic abundant	385 days

The day of *molad* is designated 1 Tishri subject to four conditions:

(i) if the *molad* falls at or after 18 hours (noon), 1 Tishri is postponed by a day;

(ii) since 1 Tishri is not allowed to fall on Sunday, Wednesday, or Friday, if the *molad* falls on those days, or if by the operation of rule (i) 1 Tishri would do so, it is postponed to Monday, Thursday, and Saturday respectively;

(iii) if the *molad* of a common year occurs on Tuesday at or after 9 hours 204 minims, 1 Tishri is postponed to Thursday, since otherwise the year would contain 356 days under rules (i) and (ii);

(iv) if the *molad* of a post-embolismic year occurs on Monday at or after 15 hours 589 minims, 1 Tishri is postponed to Tuesday, since otherwise the previous year, which must have begun after noon on a Tuesday, would contain only 382 days.

According as 1 Tishri is postponed by 0, 1, or 2 days, the preceding year is deficient, regular, or abundant.

The months are: Tishri, 30 days; Cheshvan (formerly Marcheshvan), 29 (in an abundant year 30); Kislev, 30 (in a deficient year 29); Tebet, 29; Shebat, 30; Adar, 29 days, Nisan, 30; Iyyar, 29; Sivan, 30; Tammuz, 29; Ab, 30; Ellul, 29. In embolismic years Adar has 30 days, and is followed by Ve-Adar (also called Adar Sheni) with 29 days, which takes over the festival of Purim from Adar.

There is no simple formula for converting Jewish dates to Gregorian, or vice versa; however, there is a rule for finding the

Julian equivalent of 15 Nisan (Passover as now understood), discovered by the mathematician C. F. Gauss; it is set out in the *Jewish Encyclopedia* and the *Oxford Companion to the Year*, 851–2. Since this day is always 23 weeks and 2 days before 1 Tishri, it can never fall on Monday, Wednesday, or Friday.

The Muslim calendar

Before Islam, the Arabs used a calendar of the usual lunisolar type, combined with the Seleucid era (see Chapter 7) reckoned from the Byzantine civil New Year on 1 September. The Prophet, however, broke the link between months and seasons by forbidding intercalation, relying purely on the moon and beginning the month when the new moon had been observed in the night sky by reliable witnesses; as a result, the Muslim calendar consists of 12 lunar months without correction by the sun, so that 33 Muslim years correspond within a few days to 32 Western years.

The months are called Muḥarram, Ṣafar, Rabīʿ al-ʾawwal, Rabīʿ al-ʾākhir, Jumadā ʾl-ʾūlā, Jumadā ʾl-ʾukhrā, Rajab, Shaʿbān, Ramaḍān, Shawwāl, Dhū ʾl-qaʿda, Dhū ʾl-ḥijja. Days are counted straight through; but in classical Arabic an alternative system is found in which the first of the month (in daytime) is expressed as 'one night being past' and so on up to the 14th; the 15th is 'the middle'; the 16th is 'fourteen nights remaining', and so in descending sequence to the last.

Since a calendar based on unpredictable local observations is useless for astronomical purposes, a theoretical model was devised in which the odd months are full and the even hollow, making 354 days in all. Since the synodic month is in fact a little longer than 29½ days, such a year will be short of 12 lunations by 0.367 08 day, or not quite 8 hours 48 minutes 36 seconds; to compensate, the last month of certain years is made full instead of hollow. The extra day is currently added in 11 years of a 30-year cycle, namely years 2, 5, 7, 10, 13, 16, 18, 21, 24, 26, and 29, which reduces the deficit to

0.0124 day = 17 minutes 51.36 seconds. However, this has not been uniform practice throughout the history of Islam.

Muslim years are reckoned from 16 Tammuz in the Seleucid year 933 = 16 July AD 622, the first day of the Arab year in which the Prophet departed from Mecca to Yathrib (now Medina); the Arabic for 'departure' being *hijra*, the era is called *hijrī*. However, especially in older records, the era may be counted from the previous day, 15 July; this remained the practice amongst astronomers, who began the day at noon instead of sunset. Even in Iran, where the civil calendar is solar, the epoch is still the year of the Hijra.

For means of conversion between the theoretical Muslim calendar and the Gregorian, see the *Oxford Companion to the Year*, 854–5. However, the actual Western date to which the Muslim date of a document or record corresponds may not be that given by such means, and may not be knowable at all unless (as often happens) the day of the week is given. The conversion should be adjusted to match.

The Greek calendars

Ancient Greek civilization was a system of mutually accepted variations: every city had its own dialect, its own alphabet, its own festivals, and its own laws, but recognized the others as being Greek. Naturally the calendars too varied; before Roman times these were all, at least in theory, lunisolar, each city repeating a month as it saw fit. Although there were some widespread month names, there was no unified system; from city to city the same lunation was called by different names, and the same name applied to different lunations, nor was there agreement on which lunation should begin the year or be repeated for intercalation.

In some cities the middle ten days of the month were counted separately from the first ten, and in most cities the final days were counted backwards; the last day was normally called *triakás*, the 30th, even when it was only the 29th, but at Athens the term was *hénē kaì néa*, 'old-and-new (day)'. In the Macedonian calendar, however, at least as used in Asia and Egypt after Alexander's conquests, the tale of days ran straight through from 1 to 30, the 29th being omitted in hollow months.

It is not known which cities relied on observation and which on calculation; but it is known that political or administrative convenience might lead to interference with the regular sequence of days. (The comic poet Aristophanes imagines the hungry gods turning up for their festivals only to find they are being kept on the wrong day.) As a result, not even the quantième was constant between one city and another: in 479 BC 4 Boedromion at Athens was 27 Panamos in Boeotia; in 422 BC 14 Elaphebolion at Athens was 12 Geraistios at Sparta, but a year later 25 Elaphebolion was 27 Artamihios. The report that Herostratus set fire to the temple of Artemis at Ephesus on the same day as Alexander the Great was born, if not a total fiction, may mean only that it was the same quantième in notionally corresponding months. Unless we are aided by an eclipse, it is impossible to translate even an Attic (that is, Athenian) date into a Julian, although we know the Attic calendar better than any other.

Astronomers, using a lunar calendar of alternating full and hollow days with embolismic months prescribed by a cycle, would designate the months by Attic or Macedonian names (see box) for their respectively cultural and political significance without regard to real-life calendars at Athens or in any of the lands conquered by Alexander.

The Macedonian rulers of Egypt attempted to equate their own calendar with the local lunisolar religious calendar. The task proved beyond them; by the 2nd century BC they simply applied Macedonian names to the months of the Egyptian civil year. They had an easier time in Babylonia, where the local calendar was more

Attic and Macedonian month names

The first Attic month, Hekatombaion, began at the new moon following the summer solstice; intercalation was normally achieved by repeating Poseideon. Although very little is known about the Macedonian calendar as operated in Macedon itself, it appears that the first month was Dios, beginning after the autumnal equinox; the intercalary month is unknown. In the table below, the equivalences must not be regarded as hard and fast, only as applying more often than not.

Attic	Macedonian
Hekatombaion	Loos
Metageitnion	Gorpiaios
Boedromion	Hyperberetaios
Pyanopsion	Dios
Maimakterion	Apellaios
Poseideon	Audnaios
Gamelion	Peritios
Anthesterion	Dystros
Elaphebolion	Xandikos (Xanthikos)
Mounychion	Artemisios
Thargelion	Daisios
Skirophorion	Panemos

comprehensible; the months were given Macedonian names, so that Nisanu became Artemisios. The spring New Year was retained, so that the years of the Seleucid era remained six months behind those used in other parts of the Near East; Seleucid year 1 Babylonian (the last of a Metonic cycle) began on 1 Artemisios/ Nisanu 311 BC. This calendar was retained when Babylonia fell under Parthian domination; however, from AD 17 onwards the Macedonian names are applied one Babylonian month later, Artemisios now being Aiaru instead of Nisanu. It is on this basis that the Jewish historian Josephus names Jewish months in Greek.

In Roman times many cities adopted the Julian calendar, even if with different month names and New Year: thus at Antioch October was called Hyperberetaios, and began the year (as it continued to do among Syriac-speakers) till the mid-5th century, when New Year was moved back to 1 Gorpiaios (September). Others adopted calendars on Julian principles, but with their own months; thus in the province of Asia months began on *IX Kal.* Roman, and New Year was Augustus' birthday, 23 September. This remained the ecclesiastical new year of the Eastern church, which celebrates the Conception of John the Baptist ('the Forerunner') on that date instead of the Western 24th.

The Gaulish calendar

In 1897 fragments of a calendar in Gaulish (the language spoken at the time of Caesar's invasion) were discovered at Coligny in the *département* of Ain (see Figure 20). They have been the subject of much study and more controversy; but it seems clear that the inscription itself presents a quinquennium or 5-year cycle of 62 lunar months (with embolismic months at the beginning of the first year and the middle of the third), each divided into two halves, the first of 15 days and the second of either 15 or 14. In both halves the days are counted forwards, but the sequence is interrupted by some days' changing places even between months. Quinquennia were combined into larger cycles (originally of 30 years, later, it seems, of 25), in which the first embolism of the first quinquennium was omitted.

The calendar gained gradually on the sun, though in the short term this was masked by the inevitable shortfall or excess between

20. Gaulish calendar from Coligny

lunar and solar years. In principle, the first month of the common year, Samonios, began at the winter solstice; in practice, it slipped a little earlier from cycle to cycle. On the assumption that pre-Christian Ireland had used a similar calendar, this slippage has been invoked to explain why the Irish festival of *Samhain* was fixed, once the Julian calendar was adopted, not at the winter solstice but on 1 November, which it would have reached in the mid-5th century.

The name Samonios is related to the Celtic word for 'summer'; likewise the seventh month, in principle beginning at the summer solstice, was called Giamonios, from the word for 'winter'. It appears that the names were related to the solstitial celebrations at the end of the respective seasons; the notion that *samain* (modern *Samhain*) on 1 November is the end of summer reappears in a glossary.

The Hindu calendars

Religious festivals in India are still determined by the many local calendars, most but not all either solar or lunar. Until 1957 the year directly observed in the former, and by which the latter were corrected, was not the tropical but the sidereal year, divided into 12 months each corresponding to the sun's presence in a *rāśi*; unlike the Western (and Chinese) sign of the zodiac, this was not a conventional division of the ecliptic but the actual constellation. As a result of the reform, the solar year is now tropical, and the *rāśi* is a fixed sector of the ecliptic, corresponding to that occupied by the appropriate constellation in 1957.

In solar calendars, the day begins at sunrise; the month begins according to locality with the day of the sun's entry into the new *rāśi*, the day after, or (in some cases) the day after that. The month bears the name of the *rāśi* in question except in Bengal and Tamil Nadu, which use the lunar month names.

In lunar calendars, months are divided into two 'wings' or halves, a 'bright' (waxing) half, from new moon to full, and a 'dark' (waning) half, from full moon to new. There is a standard set of names, each theoretically corresponding to a particular *rāśi*; months are named in accordance with the *rāśi* in which the sun is located at new moon. In the south, and in theoretical astronomy, the month begins with the bright half; but in the north (unless embolismic; see box) it begins with the dark half, so that until new moon the month name is one ahead of the south: thus northern Magha dark corresponds to southern Pausa dark, but both are followed by Magha bright.

The first day of each half-month is that following the new or full moon; thereafter the day is generally numbered according to the *tithi* current at sunrise, the *tithi* being the time taken by the moon to travel 12° from the sun. Occasionally a quantième has to be omitted (if a *tithi* begins after one sunrise and ends before the next)

Embolisms and suppressions

In principle, the sun enters a new *rāśi* during the course of each month; there are two qualifications.

(*a*) When the sun is in the same *rāśi* at the start of two successive months (reckoned on the southern system), the first is embolismic, with the same name as the regular month following; the north too admits an embolismic month at this point, dividing the regular month and beginning with the bright half and not like others with the dark.

(*b*) When in winter the sun enters two *rāśi*s in the same month (again as reckoned in the south), the month name corresponding to the first *rāśi* will be omitted. The same name is thus omitted in north and south.

or repeated (if a *tithi* begins before one sunrise and after the next). Irregularity of quantième does not affect the feria: Sunday the 7th is always followed by Monday even if the latter be the 9th, or the 7th again, and not the 8th.

There are numerous eras in use, mostly counted in elapsed years, and several different New Year dates within them; the most widely used is the Saka Era, used in both lunar and solar calendars, with epoch AD 78, but mention must also be made of the Vikram Samvat, used in lunar calendars, epoch 58 BC, and the Kaliyuga, a period of 432 000 solar years beginning on 18 February 3102 BC. At its conclusion the world will enter into a new age.

In addition, there are calendars based on the sidereal revolution of Jupiter, which comprises 11.862 years; five such revolutions amount to some 60 solar years.

Since 1957, for secular purposes, India has recognized two calendars: the Gregorian and the National Calendar, reckoned by the Saka Era; it uses the lunar month names and begins on 22 March (21 March in a Gregorian leap year).

Iranian calendars

Whereas the Achaemenid royal inscriptions seem to show a lunisolar calendar like the Babylonian but with different month names (and perhaps independent intercalations), the Parthian dynasty of the Arsacids (247 BC–AD 226) and the Persian Sasanians who supplanted it employed a solar calendar still used by Zoroastrians, including the Parsis of India. The year is an *annus vagus* of 365 days, comprising 12 months of 30 days (not numbered but named after the presiding spirit) and five epagomenal days named after the five groups of Gathas or Zoroastrian hymns; this calendar is said to have displaced an earlier 360-day calendar with embolisms every five or six years.

Just as the Egyptian year in principle begins at the heliacal rising of Sirius, so the theoretical Iranian Nawruz or New Year's Day is the vernal equinox; however, since the *annus vagus* ran ahead of the sun, in AD 632 Nawruz was 16 June. It is from that date that the Zoroastrian era is reckoned, in years of Yäzdegird III, the last pre-Islamic shah.

It is reported that, owing to ritual requirements, every 120 years, an embolism was made, after which the epagomenal days were postponed to follow the next regular month and thus revert to the proper time of year; however, after eight embolisms the practice fell into abeyance owing to war and turmoil, so that the extra days continued to follow the eighth month. Not all scholars believe the tradition, but the displacement of the epagomenal days under the Sasanians is a fact; it was not reversed until year 375 of Yäzdegird, when in response to the auspicious coincidence of Nawruz with the equinox on 15 March 1006 they were restored to their original position at the end of the year.

In the early 12th century the Parsis, but not the Zoroastrians of Iran, intercalated an extra month (though without moving the epagomenal days) for the sake of the coincidence; but there was no repetition. In 1746 a proposal was made to reverse the intercalation, bringing the Parsi calendar back into line with that used by the Zoroastrians of Iran. Only a minority adopted it, but their calendar (called 'Kadmi' or 'former' calendar) survives to this day besides the majority Shenshai (understood to mean 'imperial' calendar). Thus year 1374 of Yäzdegird began on 20 August 2004 in the Shensai calendar, but 21 July in the Kadmi.

In 1906 a third calendar was put forward called Fasli ('seasonal') or Bastani ('ancient'), in which Nawruz was once more the vernal equinox, and a sixth epagomenal day was added in Gregorian leap years. Most Parsis rejected the reform as contrary to religion; in Iran, by contrast, most Zoroastrians have accepted it, not least because it is closer to the civil calendar introduced by

Reza Shah in 1925 and retained after the Islamic revolution of 1979.

In this calendar, the first six months of the year (Färvärdin, Ordibehesht, Khordad, Tir, Mordad, Shährivar) have been given 31 days each to match the length of time from vernal to autumnal equinox, the next five (Mehr, Aban, Azär, Dei, Bähmän) have 30, and the last month in common years (Esfänd) has only 29. A 30th day is added, in principle according to a rule proposed in the 11th century, that the common year shall begin on the day when the sun enters Aries before noon, the leap year when it does so after noon; this is normally four years after the previous leap year but occasionally five. In practice, however, a complex cycle has been calculated in advance (see box). The era is reckoned from the vernal equinox before the Hijra (in Farsi pronounced Hejrät), 21 March AD 622.

Intercalation in the solar Hejri calendar

Intercalation is governed by a grand cycle of 2180 years, consisting in order of:

21 cycles of 128 years, comprising:
1 small cycle of 29 years;
3 small cycles of 33 years;
1 cycle of 132 years, comprising:
1 small cycle of 29 years;
2 small cycles of 33 years;
1 small cycle of 37 years.

In each small cycle there is a leap year in year 5 and every fourth year thereafter.

The current grand cycle is reckoned from AH 475 = AD 1096/7; the current 128-year cycle began in AH 1371 = AD 1992/3.

A similar leap-year principle obtains in the Bahá'í calendar, whose solar year contains 19 months of 19 days (19 being the Bahá'í mystical number), with four epagomenal days and a fifth when the equinox falls later than sunset (the start of the Bahá'í day) on 21 March; but in practice the Gregorian or Iranian leap year is often followed. The week begins on Saturday; the era is reckoned from 1844.

The Chinese calendar

The Chinese year is lunisolar, governed by the Metonic cycle, and by astronomical calculations refined many times, most notably by the Jesuit Father Adam Schall in 1644. The day begins at midnight, halfway through the first *shí* (1/12 day); if at any point within it (even if just before midnight) the conjunction of sun and moon is calculated to take place at the longitude of Beijing, it is the first of the month. The first month of the civil year (the third of the astronomical year) is the second lunation after that in which the winter solstice falls.

A 13th month is added, under current rules, in years 3, 6, 9, 11, 14, 17, and 19 of a Metonic cycle nine years behind the Western 'Golden Number' cycle; the month repeated, which must not be a winter month (1, 11, or 12), is that throughout which the sun will remain within the same sign of the zodiac, subject to the rule that the equinoxes and solstices must fall in months 2, 5, 8, and 11; if this would be breached, intercalation is postponed till after the second month of the following year.

The Chinese calendar was calculated down to 2020/1 by the Imperial Board of Mathematics; in 1931 it was prohibited by the Nationalist government of the Republic (which imposed the Gregorian calendar), but neither this prohibition, nor Communist campaigns against 'superstition', could suppress knowledge even in China proper, and did not affect Hong Kong (under British rule till 1997) or overseas Chinese (see Figure 21).

21. Chinese calendar

Besides the lunar calendar, there is a sequence of 24 'solar terms', solar half-months beginning when the sun either enters a sign of the zodiac or reaches its mid-point; various festivals are associated with these terms, above all the Qīngmíng ('Pure Bright') festival when the sun is halfway through Aries, a time for visiting the ancestors' graves. There is also a sexagenary cycle (see Chapter 7) used for year, month, day, and *shí*.

Similar calendars, but calculated for the longitudes of the local capitals, are or have been used in Korea, Japan, Vietnam, and (reckoned from full moon) Tibet.

Mesoamerican calendars

The basis of pre-Columbian Central American time-measurement, which was intimately connected with religion (see Figure 22), was a standard and unvarying 260-day cycle combining two smaller cycles of 13 and 20 days respectively.

The large cycle is often called *tzolkin* by modern scholars, from its Yucatec Maya name, but that is to privilege one language over another; in Nahuatl, for instance, spoken by the Aztecs, it was called *tonalpuhualli*. The components, by contrast, are given Spanish names, *trecena* and *veintena*, from *trece* '13' and *veinte* '20'. The days of the *trecena* were numbered from 1 to 13 (in a few places 2 to 14); those of the *veintena* had names which differed from language to language and did not necessarily mean the same thing: the 3rd, for instance, was Calli ('house') in Nahuatl but Akbal ('night') in Yucatec (see Figure 23*a*). Nevertheless, they were the same day just as Friday, *vendredi*, *sexta-feira*, *kenapura*, *Paraskeví*, and *piątek* are.

The two sequences ran concurrently: the first day of the 260-day cycle was 1 Cipactli (in Nahuatl)/Imix (in Yucatec; both meant 'alligator'), the second 2 Ehecatl/Ik ('wind'), the 13th day 13 Acatl/Ben ('cane'), the 14th 1 Ocelotl/Ix ('jaguar'), the 21st 8 Cipactli/Imix,

22. Piedra del Sol, illustrating links between calendar, astronomy, and religion. First (outer) ring: starry heavens. Second ring: Quetzalcoatl, night-god Texcatlipoca, and Pleiades. Third ring: days of *veintena*. Fourth ring: symbols of past 'suns' (eras). Centre: Tonatiula the sun-god, whose tongue recalls the Aztec sacrificial knife, flanked by eagles bringing victims' hearts.

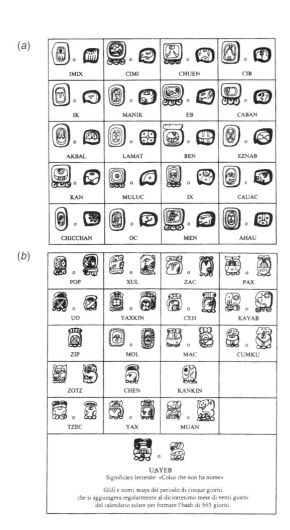

(a)

IMIX	CIMI	CHUEN	CIB
IK	MANIK	EB	CABAN
AKBAL	LAMAT	BEN	EZNAB
KAN	MULUC	IX	CAUAC
CHICCHAN	OC	MEN	AHAU

(b)

POP	XUL	ZAC	PAX
UO	YAXKIN	CEH	KAYAB
ZIP	MOL	MAC	CUMKU
ZOTZ	CHEN	KANKIN	
TZEC	YAX	MUAN	

UAYEB

Significato letterale: «Colui che non ha nome»

Glifi e nomi maya del periodo di cinque giorni
che si aggiungeva regolarmente al diciottesimo mese di venti giorni
del calendario solare per formare l'haab di 365 giorni.

23. Mayan names for (*a*) days of *veintena*; (*b*) months of solar year

and so on to the end of the cycle on 13 Xochitl ('flower')/Ahau ('lord').

Besides this cycle, each community had its solar calendar (see Figure 23b). The structure remained constant: a 365-day *annus vagus* was divided into eighteen 20-day units, called 'months' in the local languages, followed by five unlucky epagomenal days at the end; but not only was there variation both in month names and in the tale of days (some calendars counted not 1–20 but 0–19), but each community counted from whatever day it chose, so that New Year's Day in one place might not even be the first of a month in another.

Despite claims to the contrary made in nationalistic revivals of the pre-Columbian calendar, the leap year, like the wheel, was introduced by the Conquistadores; from time to time, a calendar might be replaced by a new one beginning either one day later or one month earlier, but thereafter the year would contain an invariable 365 days as before.

Each year was named according to the place in the 260-day cycle either of New Year's Day or of the 360th day; since the solar year was 105 days longer than the cycle, and $105 = 8 \times 13 + 1 = 5 \times 20 + 5$, the numerical part of the year-name rose by 1 from year to year, but the day-name advanced five places in the *veintena*. Since, moreover, $5 \times 4 = 20$, in any calendar only four day-names (known as 'year-bearers') could be used in designating a year. There were thus 52 possible year-names; when these had been exhausted, a new 'calendar round' began to great celebrations.

The Tikal Maya also recognized the *tun* of 18 months or 360 days, always ending on a day Ahau; 20 *tun* made a *katun*, of which 13 made a *may* and 20 a unit called *baktun* by modern scholars; 20 *may* or 13 *baktun* constituted a Long Count, altogether 1 872 000 days. This latter, instituted by the Olmec (probably in 355 BC), began on the completion of the last *tun* of its predecessor; within it the day was identified by the number of elapsed *baktun*, *katun*, *tun*, months, and days, by its place in the day-count, and finally by the date in the year. The current Long Count began with the ending of its predecessor on the day called 0 0 0 0 0 4 Ahau 8 Cumku, corresponding to 6 September 3114 BC, and will end on 13 0 0 0 0 4 Ahau 3 Kankin = 21 December 2012, the winter solstice.

For astronomical purposes even longer units of time were acknowledged: the *pictun* of 20 *baktun*, the *calabtun* of 20 *pictun*, the *kinchiltun* of 20 *calabtun*, and the *alautun* of 20 *kinchiltun* or 299 520 000 000 days. The date given above for the end of the current Long Count may thus be restated as 1 (*kinchiltun*) 11 (*calabtun*) 19 (*pictun*) 13 0 0 0 0 4 Ahau 3 Kankin.

Chapter 7
Marking the year

So used are we to designating years by a number in a standard era that, while we can understand that another culture may employ a different era, we are surprised when we find it does not use any era at all. Yet in the ancient world, eras, though far from unknown, were not the most characteristic means of marking the year, and when they were used were in many cases of purely local significance.

In peasant communities years are frequently identified by notable events, such as exceptionally good or bad harvests; this was still the method used under the earliest Egyptian dynasties (see Figure 24), and survives in such phrases as 'the Plague Year', denoting 1665. It has the obvious disadvantage that a year in which nothing much happened is not identifiable except in relation to one in which something did, and only over a short range.

Stating the length of time that elapsed between events is also difficult, unless a written chronicle is kept of notable occurrences. Lacking such records, the early Greeks might say that something had happened three generations ago, meaning it literally: the war was fought in my great-grandfather's time, because my grandfather told me that he had been orphaned by it. (Later historians converted generations into years, sometimes at 30 years to the generation, sometimes at 100 years to three generations, substituting a spurious exactness for long-forgotten experience.)

24. One of five fragments of the 'Palermo Stone' from Egypt, Fifth
Dynasty, *c.* 2470 BC, showing (above) detail of a list of pre-dynastic rulers
of Lower and Upper Egypt before *c.* 3000 BC, (below) events from the
reigns of First to Fifth Dynasty kings (first half of 3rd millennium BC)

These hit-and-miss methods were not good enough for more sophisticated societies, which devised means of identifying any year, however uneventful. These were by *eponym*, by *regnal year*, by *cycle*, and by *era*.

Eponyms

The *eponym*, or eponymous magistrate, was the holder of an annual office after whom the year was designated in some such formula as 'when X was [title of office]'. In Assyria this was the *limmu* or mayor of Asshur, at Athens one of the nine magistrates called archons, at Sparta one of the five 'overseers' or ephors; but the best-known example is the Roman method of dating by the two consuls of the year: *C. Iulio Caesare M. Calpurnio Bibulo consulibus* 'when Gaius Julius Caesar and Marcus Calpurnius Bibulus were consuls' (in 59 BC). It did not matter how much power, or how little, these magistrates wielded; Spartan ephors interfered at will, Athenian archons were reduced under democracy to mere administrators, the mighty consuls of republican Rome lost their power to the emperor, but still it was they, and not the latter, whose names appeared in every dated document.

Widespread as this method was in antiquity, it had three disadvantages: there was no way of identifying a future year other than as 'the so-manyeth year from now'; that since Greek cities began the year at different times, used different month-names, and did not coordinate intercalations, events that took place under one eponym in one city could not necessarily be assigned to a single eponym in another (see box); and without a list of magistrates one has no idea whether X's year precedes Y's and by how much even in one's own city, let alone abroad; such lists were indeed compiled, but were no less subject than other texts to confusion and corruption. The historian Timaeus of Tauromenium (now Taormina in Sicily) in the 3rd century BC made it his business to compare the lists of various cities, and found various discrepancies.

The difficulties of eponym dating

The inadequacy of eponym dating was illustrated when the Greek historian Thucydides needed to date the outbreak of the Peloponnesian War in spring 431 BC for a pan-Hellenic readership:

> Fourteen years had the thirty-year truce made after the capture of Euboea lasted; but in the fifteenth, when Chrysis had been priestess at Argos for 48 years, Aenesias was ephor at Sparta, and Pythodorus had two months to serve as archon at Athens, in the sixth month after the battle of Poteidaea and at the beginning of spring [a Theban force attacked Plataia]
>
> (*Histories*, book 2, chapter 2, section 1)

Thereafter Thucydides dates the events of the war by summers and winters.

Even in respect of Roman consuls, on whom the evidence after the early Republic is solid (see Figure 25), errors were not avoided in private lists.

The two consuls of AD 29 were C. Fufius Geminus and L. Rubellius Geminus; a long-standing early Christian tradition had Jesus crucified in the year of two Gemini called 'Rufus' and 'Rubellio'. In the late 4th century Epiphanius of Salamis, attempting a chronology of the Saviour's life, uses a highly inaccurate list of consuls that makes separate pairs out of 'the two Gemini' and 'Rufus and Rubellio' (though he dates the Crucifixion to neither year). No better, though different, is the list used by Prosper of Aquitaine in

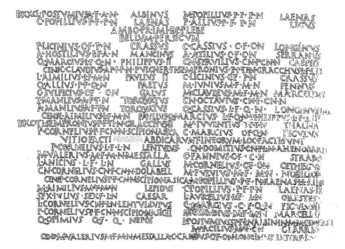

25. Part of the Fasti Capitolini (Roman consul-list)

his chronicle (AD 455) and adopted by Victorius (see Chapter 4), which puts the two Gemini one year early.

Regnal years

In the great monarchies of Egypt and the Near East the characteristic dating system was that of the king's reign. This system required lists of previous monarchs with their lengths of reigns if one was to make sense of the past (see Figure 26), and was less than convenient for the future, since one could not know how long the current reign would last even if the monarch acknowledged that it might not be for ever.

Nevertheless, the regnal year was adopted by the Emperor Justinian in AD 537 and became widespread in Europe, even outside the royal chanceries; it was not only for patriotism's sake that one learnt the Kings and Queens of England with their dates, for without them one would not know whether a document from the third year of Edward VI superseded or was superseded by one from the 23rd year

A fragment of the Turin Canon of Kings.

Part of the Table of Abydos.

Part of the Table of Saḳḳâra.

26. Fragments of Egyptian king-lists

of Henry VIII. Down to 1962 it was the official method for dating (though no longer the most frequent method for citing) United Kingdom Acts of Parliament.

But when does a regnal year begin? From Justinian onwards, it has been reckoned from the anniversary of the monarch's coming to power (see box), on whatever day it occurred. In the ancient monarchies, on the other hand, the normal principle was to count the king's regnal years from the regular New Year, irrespective of when his reign had started.

That posed the problem of the period between accession and New Year. In Sumer and Babylon this was called the beginning of the reign (or even assigned to the previous king); the modern technical

From coronation to accession

In medieval Europe monarchs often counted their years from their coronation, since it was coronation that made the king; even William the Conqueror dated his reign not from the death of Edward the Confessor, whose rightful successor he claimed to be, nor from the Battle of Hastings, when he became master of England, but from his crowning on Christmas Day 1066. However, when Henry III died on 16 November 1272, his son and heir Edward I was in the Holy Land, whence it would take him two years to return; he was proclaimed king on the 20th, from which date his regnal years were counted. Since then, 'The king's peace dies with the king' has been supplanted by 'The king is dead, long live the king' – an import from France, where the Capetian dynasty was able to transmit the crown from father to son for eight generations between 987 and 1316.

term for this reckoning is the accession-year system, since the year in which the king accedes is treated separately from the numbered years of his reign, which begin with his first New Year.

In Egypt, by contrast, the king's first year began with his accession and ended (except in the New Kingdom dynasties XVIII–XX, which used the modern anniversary system) on the last epagomenal day (see Chapter 2), so that his second and all subsequent years began on 1 Thoth. This is known as the non-accession-year system; it survived under the Macedonian rulers and the Roman emperors down to Diocletian, and was also used (when they used dates at all) by the Sasanian monarchs of Iran from the 3rd to the 7th centuries AD. It is familiar in the West as the method for stating a horse's age; it is also found in the Old Testament, but not it seems exclusively. The coexistence of accession- and non-accession-year systems, and of autumn and spring New Year (Chapter 6), makes biblical chronology a nightmare.

Egypt apart, Roman emperors did not date by their regnal years till 537; their full titles declared the number of times that they had received the annual grant of tribunician power (the legal basis for their rule), but it was not used as a date; their responses to legal enquiries, for instance, were dated by the consuls of the year. That did not prevent their subjects from counting their regnal years, if they found it convenient, but according to local principles: when St Luke dated the beginning of John the Baptist's mission to the 15th year of Tiberius, this may have been spring AD 28 to spring 29 for his Jewish Christian informant, 1 October AD 27 to 30 September 28 for Luke himself at Antioch, 29 August 28 to 28 August 29 for readers at Alexandria, and 1 January–31 December 29 for Romans.

Cycles

Faced with the multitude of eponyms with non-coincident terms of office, from the 3rd century BC Greek historians adopted the four-year cycle between successive celebrations of the Olympic Games, which had begun, according to the records, in 776 BC, and were held in the summer. This system, though confined (except at Olympia) to historical reference, could be understood by any Greek in any city, though writers still equated years in one calendar with those of another even when they began at different times, or extended their narrative over an entire campaigning season without regarding the change of year.

The Olympiad is thus an example of cyclical chronology, in which a fixed number of years are grouped into a cycle and the individual year numbered according to its position within the cycle. It is unusual in that the cycle itself is numbered. More typical is the indiction, a 15-year tax-cycle instituted (it is generally thought) in AD 312; the cycle itself is almost never numbered, but the year is regularly called the 'nth indiction', meaning the so-manyeth year of the cycle. In the late Roman Empire the indiction was very soon used in non-fiscal contexts, being more important to most people than the official dating system by consuls; in Byzantine documents it is far more reliable than the Year of the World, which as we shall see in the next section may be counted in any of several different ways.

The indiction is also found in the medieval West, partly as a residue of the late Empire, partly through dissemination in Dionysius' Easter tables (Chapter 4), but is not to be accorded the same authority; whereas in the East the indiction, reckoned at Constantinople from 1 September, enables us to determine which form of the world era is being used, in the West it is the year AD, the regnal year, or preferably both that indicate from which of several possible dates the indiction year is counted.

The most important cycle of years is the 12-year animal cycle used in central and eastern Asia, made familiar in the West by Chinese astrology. The years of the cycle are not numbered but named according to their presiding animal, the Rat, the Ox, the Tiger, the

Hare, the Dragon, the Snake, the Horse, the Sheep, the Monkey, the Fowl, the Dog, and the Pig. (There are local variations: the Buffalo may replace the Ox and in Vietnam the Cat stands in for the Hare.) In China these animals are associated with the Twelve Branches that, together with the Ten Stems, form a 60-year cycle. But this sexagenary cycle (from Latin *sexageni*, 'sixty each'), was applied not only to years, but months, days, and *shí* (double hours); in older records cyclical notation of days is more frequent than dating by quantième and month.

The 60-year cycle coexists with the *niánhào*, or 'year-name', generally translated 'era'. Before 1368 every emperor would proclaim a new era, with some auspicious name or other, at the start of his reign and again during it whenever he saw fit; thereafter each reign was its own era, and its *niánhào* was applied after death to the emperor himself. Thus the emperor in whose time the most famous Chinese porcelain was produced was not the Emperor Qianlong (or Ch'ien-lung in the older transcription), but the Qianlong Emperor, after his 'Celestial Prosperity' era.

In Japan the era (*nengō*) was not coextensive with the reign till 1868. The non-accession-year system is used, and since 1873 the Gregorian calendar: 1 Shōwa lasted from 26 December 1926, when the Shōwa emperor (Hirohito) acceded, to the 31st, 1 Heisei (the current era) from 8 January to 31 December 1989.

After the overthrow of the last Chinese emperor, a Republic (Mingguo) era was instituted that is still used on Taiwan; years are counted on an accession-year basis from 1912. That is also the epoch, being the birth-year of Kim Il Sung, for the non-accession-year Juche ('Self-Reliance') era instituted in 1997 by North Korea, named after the regime's professed principle. These may be regarded (political sentiment apart) as true eras, not limited by an individual's whim or lifespan.

Eras

The term 'era', for a chronology in which years are numbered continuously from a starting point or epoch without reverting to 1, is derived from the post-classical Latin word *aera* or *era*, properly denoting the place of an item in a numbered sequence and hence used for the serial number of the year (now called by the French term *millésime*). The chronological use originated in Spain, where years expressed in the local dating-system (for which see below) were indicated not by *anno* but by (*a*)*era* (e.g. *era mclxxiii*, as it were 'no. 1173' = AD 1135); the word was extended to mean the dating sequence itself, and then others like it. The great merits of era datings are that the intervals between events are easily calculated without the need to add up the lengths of reigns or count off magistrates from a list, and that future years may be identified as far ahead as one wishes.

The epoch of an era may be a correctly dated historical event, such as the Prophet Muhammad's *hijra* or departure from Mecca to Medina, from which the Muslim era is reckoned (see Chapter 6); but for chronological purposes it makes no difference if the date is wrong or doubtful, as in the case of the Christian era, nor if the event itself is legendary, like the accession of the Emperor Jinmu in 660 BC, from which in the ultra-nationalist period Japanese years were counted.

Eras may be reckoned either in current years, in which year 1 begins immediately after the epoch, or in elapsed years, in which it begins only when a year has already been completed. Both systems are familiar to us for stating ages: when we say that a person is in his or her 25th year we are counting current years, but when we say that the same person is 24 years old we are counting elapsed years. In era dating current years are the norm except in India; of the many Indian eras the most important is the Saka era, reckoned in elapsed years from AD 78, on which the National Calendar is based (see Chapter 6).

In Hellenistic and Roman times there were numerous local eras, commemorating political events, but few were of any significance outside the city or province concerned. These eras do not include the *ab urbe condita* reckoning from 753 BC familiar in modern writings about Rome, since Romans were not agreed on the correct date of foundation; when an event is said to have taken place so many years after the foundation, this is no more a formal dating than 'a hundred years after the Norman Conquest' would be in English.

The most important era in classical antiquity was the Seleucid era of western Asia. In 311 BC the Macedonian satrap or governor of Babylon, Seleucus, having restored himself to power by force of arms, began numbering his years of renewed office from 1 Nisanu, in that year corresponding to 3 April; when a few years later he took the title of king, he did not alter the count. His Macedonian and other Greek subjects adopted it, but, being used to Macedonian years that began in the autumn, they placed the epoch six months earlier, in late 312 instead of early 311. After his death – by which time his realm extended from Turkey to Tajikistan – the count was maintained by his successors; it remained in use throughout antiquity, was kept up by Jews (who called it 'the reckoning of contracts') till the Renaissance (even longer in Yemen), and survived amongst Nestorian Christians till the second half of the 20th century, when, styling themselves the Assyrian Church of the East, they adopted an era with epoch 1 April 4750 BC, based on a surmised foundation date of Asshur.

Other eras, such as that of Provincia Arabia (epoch 22 March AD 106), were more localized and mostly short-lived; an exception in the latter respect is the Hispanic era, with epoch 1 January 38 BC. This is traditionally associated with Augustus, for no clear historical reason, but may commemorate the beginning of Roman conquest in the Pyrenaean region where the earliest (but contested) examples of this reckoning have been found. The era is indubitably attested from the late 4th century; it was used in Visigothic Spain (except for the easternmost portions, where it appears only after the *Reconquista*), and lasted in official use till the later Middle Ages: in Aragon till 1350, in Castile till 1383, and in Portugal till 1422.

World eras

The era that would ultimately displace the Seleucid era amongst Jews was a world era, that is one reckoned from the creation of the world; for this purpose they adopted the epoch already used for calendrical calculation, 3761 BC (see Chapter 6). The Jewish year from 16 September 2004 to 3 October 2005 is thus AM 5763, often (especially in Hebrew) written '763; AM, standing for *annus mundi*, or 'year of the world', is the conventional qualification for a year in any world era, including those devised by Christians. The basis of such eras was the chronology of the Old Testament, which is far from simple and is also considerably shorter in the Hebrew text and St Jerome's Latin version than in the Greek translation known as the Septuagint. World eras were mainly developed by Greek-speakers, beginning with Sextus Julius Africanus (*c.* 221), who placed the conception of Christ on 25 March and made it the first day of AM 5501. This is commonly taken to be 2–1 BC (though not all his datings fit). Nearly a century later, Eusebius of Caesarea dated Creation to 5200 BC, Christ being born in AM 5199; however, he preferred to call this 'year of Abraham 2015'.

Disseminated through Jerome's translation of his *Chronicle*, Eusebius' calculation became the standard theory in the West till Bede, using the Latin Bible, reduced the period between Creation and Nativity to 3952 years. Other Greek-speakers, however, preferred the higher interval of Africanus, or one close to it, but adjusted so that the Creation should take place on a Sunday; the most favoured was the era of Annianus (early 5th century), in which the Creation took place on Sunday, 29 Phamenoth = 25 March 5492 BC, and the Incarnation, meaning the Conception of Jesus Christ, on Monday, 29 Phamenoth AM 5501 = 25 March AD 9.

However, although reckoning the year from the anniversary of Creation was theologically attractive, in practical life it was inconvenient; the epoch was therefore adjusted to the civil New Year before Creation, 1 Thoth/29 August 5493 BC. This caused Incarnation and Nativity to fall in different years: rather than redesignate the year of Nativity 5502, the Alexandrians reassigned the Incarnation to AM 5500, which had the advantage of placing it on a Sunday, 25 March AD 8; the AM 5501 in which the Nativity supposedly took place now began on 29 August AD 8. This became the first year in an era still used in Ethiopia, where the Year of Grace 2000 will begin on 30 August Old Style = 12 September 2007.

The 7th-century *Chronicon Paschale* (so called because it began with an account of Easter reckoning) opted for 25 March 5509 BC. Later Byzantines, however, preferred to defer the Creation till the beginning of the civil year on 1 September; an unsuccessful alternative was 25 March 5508. In Russia the year of Creation was the regular dating system, reckoned originally from 1 March 5508 (less often 5509) BC but by the later 14th century from 1 September 5509 BC, till by decree of Peter the Great 31 December AM 7208 was followed by 1 January 1700 Old Style.

Perpetuated reigns

Some eras developed out of regnal years continued after the death of the monarch: as we saw in Chapter 6, the Zoroastrian era commemorates Shah Yäzdegird III, whose first regnal year began on 16 June AD 632. Several such eras were created by astronomers, who found continuous numeration helpful; one is the era of Nabonassar, reckoned in Egyptian *anni vagi* from 1 Thoth = 26 February 747 BC, the first year (on the Egyptian reckoning) of the Babylonian king from whose time onwards astronomical records were preserved. Another is the era of Diocletian.

When Augustus (as he later became) conquered Egypt in 30 BC, he ruled it as king through a viceroy or prefect, outside the general provincial system, counting his years on the established non-accession system. His successors followed suit until Diocletian, at the end of the 3rd century AD, integrated Egypt into his reformed provincial structure and introduced consular dating. That was highly inconvenient for astronomers, who would need to keep lists of consuls in order to understand their own observation-records; instead, they continued to count by Diocletian's regnal years, of which the first was 284/5, even after his abdication in 305. This was the method used to designate years in Alexandrian Easter tables; it spread to general dating purposes, and is still the favoured era of the Coptic church. However, since Diocletian, in his last years of power, had unleashed the Great Persecution against the Church, from the 7th century the era was renamed that of the Martyrs. After year 532 of the Martyrs (= AD 815/16) years are sometimes numbered over again in Paschal cycles of 532 years, so that (for instance) year 257 may be not 540/1 but 1072/3 or 1604/5.

Christian era

The odiousness of the persecutor's name was also the reason given by Dionysius Exiguus for replacing, in his Easter table, the era of Diocletian with that of the Incarnation, 'so that the beginning of our hope might be better known to us and the cause of human restoration, that is the passion of our Redeemer, might shine forth more clearly'. The Incarnation is not the Passion; but Dionysius was brushing aside his predecessor Victorius, who had designated the years in his table by an era of the Passion reckoned from AD 28, his compatriot Prosper's inaccurate date for the two Gemini. (This was not the only Passion era known: at Rome in Bede's day years were counted from AD 34, or perhaps 33; other dates are found in the East.)

Dionysius treats his Incarnation date as unproblematic and uncontroversial, neither explaining how it is known nor claiming it as his own discovery. Since most earlier writers had dated the Incarnation to 2 BC, this has been difficult to explain: one theory requires him to misread or misrepresent the Olympiad date of Diocletian's accession in Eusebius' *Chronicle*, compiled in the late 3rd century, or its translation by Jerome; however, since the Nativity in AD 1 is already found in a calendar written in 354, another scholar shifts the blame to Eusebius, supposing a miscalculation in the Easter table that we know him to have written.

Another suggestion is that Dionysius deliberately fudged his figures in order that leap years should continue to be divisible by 4, as in the Alexandrian tables; for although the leap day had been added in the previous year, it was in the exact multiples, such as year 244 of Diocletian, that it affected the Easter calculations. It was and is convenient that year 248 of Diocletian should be 532 of the Incarnation, rather than 531 or 533. The Church historian Socrates, translating into Greek the report that the emperor Valens began his reign on *V Kal. Mart.*, rendered it in the normal way as 25 February without realizing that the year in question was a leap year, so that the correct date was the 26th. Had he, like us, known the year as 364, he could have seen the fact at once.

Nevertheless, Dionysius' date shares with the 2 BC it displaced the defect of being incompatible with both Gospel narratives, for St Matthew's story of the Magi and the Holy Innocents requires the Nativity to have taken place at least two years before the death of Herod the Great at Passover 4 BC, and St Luke's narrative places it in AD 6, when 'Cyrenius', that is to say P. Sulpicius Quirinius, was incorporating Judaea into the Roman province of Syria. No solution of the problem has yet satisfied either believers or non-believers in the literal truth of the Bible.

The year of the Incarnation

When preachers say on Christmas Day that Christ was born so many years ago, they always give the number of the current year, implying that the Nativity took place on 25 December 1 BC; that was also the view of those churches and orders that counted the era from that date (see below). By contrast, although this is the first year in Dionysius' 19-year cycle, Bede, following Irish sources, took him to have put the Incarnation in a year whose characteristics match the second year of his cycle, AD 1; this is more compatible with the preference for current over elapsed years, though the computist of 243 had devised an elapsed-year era of the Exodus. Dionysius himself is unlikely to have given the matter any thought.

The spread of AD dating

Dionysius' Incarnation era, like Victorius' Passion era, was originally devised for the sake of Easter tables; a few authors use it for relative chronology, typically in conjunction with the incompatible chronology of Eusebius. However, the habit of writing annals, or brief records of a year's events, in the blank spaces of Easter tables encouraged a closer association between era-date and year; this was particularly congenial to Irish and English monks, for whom the Emperor was a foreign potentate and whose countries were divided among numerous kings and kinglets.

Although the prevalent means of identifying the year in Ireland, at least in monastic writings, was by the feria and lune of 1 January, we find explicit dating by Victorius' Passion era as early as 658. In Northumbria by the late 7th century Dionysius' Easter reckoning prevailed over Victorius'; it was therefore Dionysius' era that Willibrord, the apostle of the Frisians, employed when he noted in his calendar that he had crossed the sea to Francia 'in the 690th year from Christ's Incarnation', had been ordained bishop in 695, and was now living in 728.

The decisive moment, however, was Bede's decision to use this reckoning in his *Ecclesiastical History of the English People*, rather than the world era of his chronicle; the *History*, an instant classic, brought Incarnation dating to the attention of Continental readers, who in due course began to adopt it for themselves. Although alternative epochs of 22 BC (Abbo of Fleury in the 10th century) and

23 BC (Marianus Scottus of Fulda in the 11th) were proposed in order to salvage the Western tradition that the Crucifixion took place on 25 March, which was *luna XIV* in AD 12, and Gerlandus of Lotharingia in the 11th century adapted the Alexandrian Incarnation era to the Julian calendar by subtracting seven years from the date AD, the Dionysian era prevailed, ousting even the deep-rooted Hispanic era, to become the world-wide standard even outside Christendom.

Dating 'before Christ'

The Christian era is the only era in which dates before the epoch are regularly identified as such; if occasional instances in the Middle Ages are still comparable with casual references to events so many years before the foundation of Rome or the Hispanic era, since the 18th century it has been normal to count 'years before Christ' on the same footing as 'years of our Lord'. The main resistance came from German historians of ancient Rome, who preferred to canonize the 'Varronian' date for the city's foundation, and switch to the Christian era only from the epoch onwards, so that 753 was followed by 1; this usage is now obsolete.

Astronomical dating

Whereas in normal usage AD 1 is preceded by 1 BC, in astronomical reckoning the year 1 (unlabelled) is preceded by year 0, and that in turn by –1, corresponding to 2 BC; correspondingly 45 BC is –44, 100 BC is –99, and so on. This not only assists calculation (from –7 to 3 is $3 - (-7)$ years $= 3 + 7$ years $= 10$ years), but makes all years divisible by 4 leap years; in the normal reckoning this applies only to years AD, those BC being leap if of the form $4n + 1$.

Ideological content of eras

Although a regnal year may send a message at a time of political contention, it is eras that are the most obviously ideological form of chronology. To the many examples already seen may be added the turmoil caused in Iran when on 24 Esfänd 1354 solar Hejri (the Farsi pronunciation of *hijrī*), corresponding to 14 March 1976, Mohammad Reza Shah decreed a new Shahänshahi ('Imperial') era, reckoned from Cyrus the Great's accession to the Persian throne in 559 BC, to begin a week later (1354 being a leap year) at Nawruz 2535.

This, one of many attempts at associating the dynasty with the glorious Achaemenids of ancient times, was received by the people as an affront to Islam. A Western reader may conceive some faint idea of the indignation aroused by imagining that Mussolini, instead of instituting a Fascist Era with epoch 29 October 1922 to be used concurrently with the Christian, had replaced the Christian era with that of Rome, so that 1923 had become 2676. Popular protest forced restoration of the Hejri era from 5 Shährivar 1357 (27 August 1978).

The Christian era is too well established to be challenged for its religious origin; in China indeed, where Christianity has never been more than a minority religion, it was made official by the anti-religious Communists. However, the name has come under attack; whereas Muslims freely speak of the *mīlādī* or 'Nativity' year, Continental secularists prefer to call the era simply 'ours' (*notre ère, unsere Zeit*), and amongst English-speakers the term 'Common Era', already standard in Jewish usage (compare Hebrew *ha-sefirah*, 'the count'), has become widespread in American academic writing. Even some Christians have accepted it, whether in an anti-proselytizing spirit or because there are no grounds for believing the era's epoch to be the true date of the event that it commemorates. Nevertheless, if it does not commemorate the birth of Christ, it has no business to exist at all, for no other event of world-historical significance took place in either 1 BC or AD 1.

Beginning of the year

If Incarnation and Nativity are to fall in the same year, it must begin no later than 25 March; but this date is impossible for a computistic year, since Easter may precede it. Yet Dionysius' lunar regulars presuppose a year beginning in September as at Byzantium (it was Bede who recalculated them from January); if forced to specify, he might have stated that his epochal year ran from 1 September to 31 August, incorporating the Incarnation, from which he counts, but not the Nativity, from which he does not.

His Western readers, however, took some time to recognize the difference between Incarnation and Nativity. It was quite frequent for years to be reckoned, not from 1 January AD 1 – a date disliked by the Church on account of the pagan festivities it had failed to suppress – but from seven days previously, 25 December 1 BC, the supposed date of the Nativity. This, despite Bede, was the practice in Anglo-Saxon England, and long remained in use in Benedictine monasteries; but it was ultimately supplanted by the rival principle of counting from the Incarnation proper on 25 March, the Annunciation or Lady Day. In the late 10th century, we find in parts of southern France and northern Italy an epoch of 25 March 1 BC, resulting in a millésime 1 higher than in the modern reckoning till 31 December; this fell out of favour except in Pisa, for which reason it is known as the *calculus Pisanus*. More widespread was Annunciation in AD 1, with a millésime 1 lower than the modern between 1 January and 24 March; this was characteristic of Florence and England, for which reason it is known as the *stilus Florentinus*, or the 'custom of the English church' (*consuetudo ecclesiae Anglicanae*).

Pisa and Florence retained their respective usages down to 1749, before being ordered to count from 1 January by Grand Duke Leopold of Tuscany; the English style was reformed by Act of

Parliament in 1751 (Scotland had used 1 January since 1600). Venice preferred to count from the beginning of the Incarnation month, that is 1 March AD 1, and continued to do so in official documents till the suppression of the Republic in 1797. If this *mos Venetus* was more convenient than changing the millésime within a month, the French custom, *mos Gallicus*, of beginning the year at Easter was less so: but even after the royal ordinance abolishing it in 1564 local resistance prolonged its use in some parts of the country (in the Beauvaisis till 1580).

Exact study of documents has shown that the medieval dates for the change of millésime varied within as well as between countries to an even greater extent than is stated in reference books. Nevertheless, throughout Europe west of the Byzantine Empire, 'New Year' and its equivalents in other languages regularly meant 1 January even before the adoption of the Modern Style, as counting from that day is known.

Hybrid systems

Some Christian chronologies state the years of their eras according to the 532-year Paschal cycle: in Georgia from the 9th to the 19th century, dates were given in years of the *kronik'oni*, a Paschal cycle reckoned from AD 781 or 1313, respectively the 13th or 14th from Creation in 5604 BC. Coptic years of the Martyrs may also be reduced to years of a Paschal cycle (see p. 122).

Designation by characteristics

As we have seen, Irish monks commonly designated years by the lune and feria of 1 January; the lune might be taken from the Latercus, Victorius, or Dionysius, depending on the custom of the house. Designation by characteristics, in this case the place of the year in the *trecena* and *veintena* of the 260-day cycle, was also the norm for Mesoamerican solar years.

Julian Period, Julian Day

The work of making chronological sense out of ancient data was begun by the great polymath Joseph Justus Scaliger in his *De emendatione temporum* (1583), with the help of a new dating method, the Julian Period. This was a cycle of 7980 years, combining the 19-year Golden Number cycle, the 28-year solar cycle, and the 15-year cycle of indictions; since the next 15th indiction ending an Easter cycle in the unreformed calendar was 3267, Scaliger made that year JP 7980, so that JP 1 was 4713 BC. For any year BC, the year JP is obtained by subtraction from 4714, for any year AD by adding 4713; its place in the cycles is the remainder to 19, 28, and 15 respectively. Thus 1583 was JP 6296, Golden Number 7, solar cycle 24, indiction 11.

Unfortunately, Pope Gregory's reform (which Scaliger, as a Protestant, of course opposed) had just abolished the Paschal cycle, and the indiction was of no practical use; nevertheless, astronomers have found the epoch of service as the basis for a continuous count of Julian Days, which is counted in elapsed days from noon on Monday, 1 January JP 1, also written –4712 I 1; the 24 hours from then till noon on 2 January JP 1 are thus JD 0. When the Julian Day is followed by a decimal representing the fraction of the day elapsed since the preceding noon, it becomes the Julian Date. In order to respect the midnight start to the day adopted in 1925, and to avoid high numbers, the Modified Julian Date or MJD is often used; this is the Julian Date minus 2 400 000.5. For example, 6 a.m. on 31 March 2004 is MJD 53 095.25, corresponding to Julian Date 2 453 095.75.

The Julian Period is not to be confused with the Julian Year, counted from the introduction of the Julian calendar in 45 BC = JP 4669, mentioned by Censorinus in AD 238 and used by some early modern writers for discussions of New Testament chronology. It would make an excellent secular – and politically uncontentious – substitute for the Christian era, being related to the calendar rather than any external event, but for the inconvenience that leap years are of the form $4n + 1$ instead of exact multiples of 4, and the even greater inconvenience that a change to the common era of the human race would cause confusion and expense even in those countries that officially use a different reckoning.

Appendix A

The Egyptian calendar

Month		1st day of month in year				
		BC	BC	BC	BC	AD
Old name	New name	1322/1	592/1	238/7	26/5	139/40
Flood						
1st month	Thoth	20 July	18 Jan.	22 Oct.	30 Aug.	20 July
2nd month	Phaophi	19 Aug.	17 Feb.	21 Nov.	29 Sept.	19 Aug.
3rd month	Hathyr	18 Sept.	19 Mar.	21 Dec.	29 Oct.	18 Sept.
4th month	Choiak	18 Oct.	18 Apr.	20 Jan.	28 Nov.	18 Oct.
Winter						
1st month	Tybi	17 Nov.	18 May	19 Feb.*	28 Dec.	17 Nov.
2nd month	Mecheir	17 Dec.	17 June	20 Mar.	27 Jan.	17 Dec.
3rd month	Phamenoth	16 Jan.	17 July	19 Apr.	26 Feb.*	16 Jan.
4th month	Pharmouthi	15 Feb.*	16 Aug.	19 May	27 Mar.	15 Feb.*
Summer						
1st month	Pachon	16 Mar.	15 Sept.	18 June	26 Apr.	16 Mar.
2nd month	Payni	15 Apr.	15 Oct.	18 July	26 May	15 Apr.
3rd month	Epeiph	15 May	14 Nov.	17 Aug.	25 June	15 May
4th month	Mesore	14 June	14 Dec.	16 Sept.	25 July	14 June
'Days upon the year'		14 July	13 Jan.	16 Oct.	24 Aug.	14 July
Next year began		19 July	18 Jan.	21 Oct.	29 Aug.	19 July

*29-day month

NB. The new month names are first attested for the solar calendar in the 6th century BC; they were taken from the lunisolar ritual calendar, which was so regulated that Thoth lunar always began in Thoth solar.

Appendix B

The Alexandrian Easter

As stated in Chapter 4, the Easter reckoning that became standard at both Rome and Constantinople was an adaptation to the Roman calendar of Alexandrian calculations in which a lunar Paschal calendar was mapped onto the civil year as reformed by Augustus (see the Alexandrian calendar below).

The lunar calendar was a notional Jewish calendar not actually used by Jews whether at Alexandria or elsewhere, in which no notice was taken of the rules restricting the feria of 1 Tishri. It comprised 12 months, alternately full and hollow, each conceived as beginning in the corresponding solar month, as in the Egyptian religious calendar, rather than ending in it as in the West. In order that the year should begin no earlier than 15 Thoth, an embolismic month of 30 days was added at the end of years 2, 5, 7, 10, 13, 16, 18 in the cycle. The civil leap day, corresponding to 29 August in a Julian pre-leap year, and occurring either four or five times within the cycle, was not given its own lune; since $19 \times 365 = 6935$ days and $19 \times 354 + 7 \times 30 = 6936$, in the last year of the cycle there was a *saltus* at the end of the 11th lunar month, so that lunar age 30 was reached on 5 Epagomenon (28 August). The last lunar month then ran from 1 to 29 Thoth, the first solar month of the new cycle, and the first lunar month on 30 Thoth. Since the lune of 5 Epagomenon

132

was taken as the epact of the following year, the cycle began each time with an epact variously designated 30, 29 (since the *saltus* made the 11th lunar month hollow), or 0.

In order to find the feria of any given day, Alexandrian astrologers had devised a two-part algorithm. The first part consisted in dividing the year of Diocletian by 4, ignoring the remainder, adding the quotient to the whole and also the parameter 2, and then taking remainder of the total to 7; if there is no remainder, call the answer 7. This process yielded the 'days of the gods' for the year (the gods being those who ruled the planets); from these the feria of the given date was found by adding to them 2 for every month up to and including it, plus the quantième, and taking the remainder to 7.

However, Christians noticed that the 'days of the gods' corresponded to the feria of 1 Thoth, and therefore of 1 Pharmouthi (27 March), in a week counted from Wednesday instead of Sunday. They therefore adopted it, without change of name, as the basis for finding the Sunday after *luna XIV*; if, for instance, *luna XIV* fell on 3 Pharmouthi (29 March) and there were 7 days of the gods, they knew that 1 Pharmouthi was a Tuesday, that 3 Pharmouthi was a Thursday, and therefore that Easter Day would be on the 6th (1 April).

Since the feria on which a given date falls will be one later every year except when further advanced by an intervening leap day, and since there are 7 feriae in the week and 4 years in the leap-year cycle, the full pattern of correspondences recurs after $7 \times 4 = 28$ years (the solar cycle). Since there are 19 possible dates for the *luna XIV*, the full pattern of Easter dates recurs after $28 \times 19 = 532$ years; this is known as the Paschal cycle. However, although the Alexandrians knew this, they generally drew up their tables to cover only 5 Metonic cycles, making 95 years, partly because there were still expectations that the world would end in its 6000th year, which was expected about AD 500, and partly because in the Alexandrian calendar the same date continues to fall on the same feria at 95-year intervals until it reaches a leap year, when it is one feria earlier. (In the Julian calendar, the same is true before the leap day, but from leap day onwards, and hence at Eastertide, the feriae of the later year will be the same; on the other hand, if the *earlier* year is a leap year, the feriae in the later year will match it before leap day and then diverge.)

The Alexandrian Calendar

1 Thoth	29 (30*) August	1 January	6 (5†) Tybi
1 Phaophi	28 (29*) September	1 February	7 (6†) Mecheir
1 Hathyr	28 (29*) October	1 March	5 Phamenoth
1 Choiak	27 (28*) November	1 April	6 Pharmouthi
1 Tybi	27 (28*) December	1 May	6 Pachon
1 Mecheir	26 (27†) January	1 June	7 Payni
1 Phamenoth	25 (26†) February	1 July	7 Epeiph
1 Pharmouthi	27 March	1 August	8 Mesore
1 Pachon	26 April	1 September	4 (3*) Thoth
1 Payni	26 May	1 October	4 (3*) Phaophi
1 Epeiph	25 June	1 November	5 (4) Hathyr
1 Mesore	25 July	1 December	5 (4) Choiak
Epagomenai	24–8 (24–9*) August		

* in Julian pre-leap year

† in Julian leap year

Further reading

A fuller treatment of subjects discussed in this book will be found in Bonnie Blackburn and Leofranc Holford-Strevens, *The Oxford Companion to the Year* (Oxford, 1999). The bibliography in this work and others listed below should be consulted.

Much information on calendars in general is provided by E. G. Richards, *Mapping Time: The Calendar and its History* (Oxford, 1998) and Nachum Dershowitz and Edward M. Reingold, *Calendrical Calculations* (Cambridge, 1997); the calculations in the latter book are devised for execution on the computer. Duncan Steel, *Marking Time: The Epic Quest to Invent the Perfect Calendar* (New York, 2000), is strong on matters astronomical. Less mathematically demanding is David Ewing Duncan, *The Calendar: The 5000-Year Struggle to Align the Clock and the Heavens – and What Happened to the Missing Ten Days* (London, 1998).

Alan E. Samuel, *Greek and Roman Chronology: Calendars and Years in Classical Antiquity* (Munich, 1972) is an invaluable resource; a broader study is E. J. Bickerman, *Chronology of the Ancient World* (London, 1968 and later editions).

Byzantine and other Eastern Christian chronology is treated by Victor Grumel, *La Chronologie* (Paris, 1958). The Insular 84-year cycle was first brought to light by Daniel McCarthy and Dáibhí Ó Cróinín, 'The

'Lost' Irish 84-Year Easter Table Recovered', *Peritia*, 6–7 (1987–8), 225–42, reprinted in Ó Cróinín, *Early Irish History and Chronology* (Dublin, 2003), 58–75; an improved account is given by McCarthy, 'Easter Principles and a Lunar Cycle Used by Fifth Century Christian Communities in the British Isles', *Journal for the History of Astronomy*, 14 (1993), 204–24.

Easter apart, the Christian liturgical year falls outside the scope of this study; among the many works dealing with it is Thomas J. Talley, *The Origins of the Liturgical Year*, 2nd edn. (Collegeville, MN, 1991).

On the week, see Eviatar Zerubavel, *The Seven-Day Cycle: The History and Meaning of the Week* (Chicago, 1989).

The Jewish and Muslim calendars are discussed in the standard reference works of the respective religions; Sherrard Beaumont Burnaby, *Elements of the Jewish and Muhammadan Calendars* (London, 1901) describes their mathematical bases in detail and gives extensive tables of equivalences. On the Jewish calendar's transition from observed to calculated, see Sacha Stern, *Calendar and Community: A History of the Jewish Calendar, 2nd Century BCE–10th Century CE* (Oxford, 2001). Complete conversion tables for the Muslim calendar will be found in G. S. P. Freeman-Grenville, *The Islamic and Christian Calendars, AD 622–2222 (AH 1–1650)*, 3rd edn. (Reading, 1995).

On the Chinese calendar, see Pierre Hoang, *A Notice of the Chinese Calendar and a Concordance with the European Calendar* ('Zi-Ka-Wei near Shanghai' = Xijiahui, 1900). There is a detailed study of the Coligny fragments by Garrett Olmsted, *The Gaulish Calendar* (Bonn, 1992). In India, the annual Rashtriya Panchang sets out calendars for the coming year according to the chief systems; for those in operation before 1957, see Robert Sewell and Śankara Bâlkrishna Dîkshit, *The Indian Calendar* (London, 1896).

For Mesoamerican calendars, see Munro S. Edmonson, *The Book of the*

Year: Middle American Calendrical Systems (Salt Lake City, 1988);
Alfonso Caso, *Los calendarios prehispánicos* (Mexico City, 1967);
Anthony Aveni, *Skywatchers*, 3rd edn. (Austin, TX, 2001).

Aveni is also the author of the more general study *Empires of Time: Calendars, Clocks, and Cultures* (London, 1990; paperback 2000). See too G. J. Whitrow, *Time in History: Views of Time from Prehistory to the Present Day* (London, 1988); Arno Borst, *The Ordering of Time: From the Ancient Computus to the Modern Computer*, trans. Andrew Winnard (Cambridge, 1993). Kristen Lippincott (ed.), *The Story of Time* (London, 1999), is richly illustrated and wide-ranging, as is Émile Biémont, *Rythmes du temps: astronomie et calendaires* (Paris and Brussels, 2000). J. T. Fraser (ed.), *The Voices of Time: A Cooperative Survey of Man's Views of Time as Expressed by the Sciences and by the Humanitie*s, 2nd edn. (Amherst, MA, 1981) is also of interest.

Accessions and deaths of rulers, and useful information on the beginning of the year and the adoption of the Gregorian calendar in various countries, as well as Easter dates, will be found in Adriano Cappelli, *Cronologia, cronografia e calendario perpetuo*, 7th edn. rev. Marino Viganò (Milan, 1998). Tables of several calendars are given by Frank Parise, *The Book of Calendars* (New York, 1982), sometimes with excessive confidence.

Marie-Clotilde Hubert (ed.), *Construire le temps: normes et usages chronologiques du moyen âge à l'époque contemporaine* (Paris and Geneva, 2000), reprints articles published the previous year in the *Bibliothèque de l'École des chartes*, several of which have proved useful for the present work.

Glossary

accession-year system: regnal-year count from which the period between accession and New Year is excluded

***annus vagus* (plural *anni vagi*)**: year in a calendar without *intercalation*

apparent solar time: time measured by the sun as observed, time by the sundial; opposite of *mean solar time*

artificial day: period of daylight

civil day: calendrical day as defined by law or custom

common year: year without *intercalation*

concurrent (earlier *concurrents* or *concurrent days*): number expressing relation of year and week; understood in the Western calendar as the *feria* of 24 March

current years: a count including the year in progress

decemnovenal cycle: the *Metonic cycle* used in Alexandrian and Western Easter tables

ekeweek: intercalary week

elapsed years: a count of completed years only

embolism (adjective *embolismic*): same as *intercalation*, but particularly used of additional month

epact: the *lune* of a given day; also that of a defined day as expressing relation of given solar year to the lunar calendar

epagomenal days: days not incorporated in a month

epoch: date from which an era is reckoned

equinox: day on which night and daytime are each 12 hours

feria (plural *feriae*): day of the week

full month: in a *lunar calendar*, a month of 30 days; opposite of *hollow month*

Golden Number: place of year in *decemnovenal cycle*

hollow month: in a *lunar calendar*, a month of 29 days; opposite of *full month*

indiction: a cycle of 15 years instituted (probably) in AD 312/13, or the place of a given year within the cycle (e.g. '5th indiction' = the 5th year of the cycle)

intercalation (adjective *intercalary*): addition of extra day, week, or month to year; also called *embolism*

lunar calendar: calendar based on revolution of moon round earth

lunar cycle: *Metonic cycle*, specifically that used in Byzantine Easter tables

lunation: period from new moon to new moon; also called *synodic month*

lune: day of lunar month

lunisolar calendar: calendar adjusted by *embolisms* to keep in line with the seasons

mean solar time: time by the sun adjusted on assumption of equidistance between earth and sun throughout the year, time by the clock; opposite of *apparent solar time*

Metonic cycle: cycle of 19 lunar years with 7 *embolisms*

millésime: serial number of the year

Modern Style: reckoning from 1 January

natural day: period of earth's rotation on axis, 24 hours; day in reformed calendar on which an event would have fallen but for the reform

New Style: Gregorian calendar in its civil aspect

nominal day: day in reformed calendar corresponding to same date in unreformed

non-accession-year system: regnal-year count in which the period between accession and New Year is year 1

Old Style: Julian calendar

quantième: day of the month

saltus (**in full *saltus lunae***): leap on successive days from one *lune* to the next but one

sidereal year: period between two appearances of the sun in the same position relative to the stars

solar calendar: calendar based on earth's revolution round sun

solar cycle: cycle of 28 years after which relation of year to the week and the intercalary cycle repeats

Sunday Letter: letter of A–G cycle written against day in calendar, or letter in a given year corresponding to Sunday

synodic month: same as *lunation*

tale of days: the method of counting the *quantième*

tropical year: period from *vernal equinox* to vernal equinox; astronomically, of one complete revolution of the sun's mean longitude with respect to the dynamical equinox

vernal equinox: equinox of spring

Index